Estimating Centrifugal Compressor Performance

Gulf Publishing Company
Book Division
Houston, London, Paris, Tokyo

Estimating Centrifugal Compressor Performance

Process Compressor Technology
Volume 1

Ronald P. Lapina, P.E.

Principal Mechanical Engineer
Procon International, Inc.
Houston Operations
A Subsidiary of UOP

to my wife, Rachel, and Chris and Kim

PROCESS COMPRESSOR TECHNOLOGY
Volume 1
Estimating Centrifugal Compressor Performance

Library of Congress Cataloging in Publication Data
Lapina, Ronald P.
Estimating centrifugal compressor performance.
Bibliography: v. 1, p.
Includes index.
Contents: v.1. Process compressor technology.
1. Compressors—Performance. I. Title.
TJ990.L367 621.5'1 82-3124
ISBN 0-87201-101-1 (v. 1) AACR2

Acknowledgments

I would like to thank the many contributors to this book, namely American Society of Heating, Refrigerating and Air-Conditioning Engineers, Inc.; *Chemical Engineering*, McGraw-Hill Publications Company; Compressed Air and Gas Institute; Dresser Clark Division, Dresser Industries, Inc.; Elliott Company; Gas Processors Suppliers Association; Ingersoll-Rand Company; Pratt and Whitney Aircraft Group; Roots Operations, Dresser Industries, Inc.; Texas A&M University.

Much appreciation is due Elliott Company for their permission to paraphrase sections of the Elliott *Compressor Refresher* and to use the many sketches that appear in Chapter 1.

Much appreciation is also due the Gas Processors Suppliers Association for their permission to reproduce the several tables and figures from the GPSA *Engineering Data Book* and *SI Engineering Data Book* found mainly in the appendixes.

I further extend my gratitude to Steve LaGrone and Sam Ashy of Procon International, Inc., Houston Operations, a Subsidiary of UOP, who saw the need for this book and encouraged me to write it. I am also grateful to all my friends at Procon International for their encouragement.

And, last, I extend thanks to my editor, Nancy Pierson, for making this an enjoyable and rewarding experience.

Contents

Preface

When new plants are designed or existing plants expanded, the process engineer uses the tools and experience available to optimize plant efficiency. Often, however, the process is designed without sufficient consideration of available compression equipment. Once the process design is complete, or nearly so, the mechanical engineer will search the market for compression equipment to satisfy process requirements. At this stage, it becomes difficult, if not impossible, to vary process conditions in order to optimize compressor selection. A loss in overall plant optimization will result if the efficiency of the compression equipment has to be compromised to meet process demands.

During the process design phase, the process engineer must also perform utility balances to determine which type of power will be used to drive the compression equipment. In order to perform a utility balance, the engineer must know how to predict power requirements.

The purpose of this book is to give the process and mechanical engineer a method for estimating centrifugal compressor performance. The estimating procedure presented will enable the engineer to determine the feasibility of satisfying desired process conditions and will therefore prove useful during the initial stages of process design.

Chapter 1 describes the centrifugal compressor, its aerodynamic components and applications. This chapter may be basic for some; however, an understanding of how the centrifugal compressor works is necessary to develop a working knowledge of the subjects that follow.

Chapter 2 is a review of thermodynamics as it applies to gas compression. This chapter examines the basic applicable thermodynamic relationships and from these derives the necessary equations for compressor estimates. The adiabatic and

A two-body, 16,000 ICFM, 5,900 HP, multi-stage centrifugal compressor train driven by electric motor and gear. Photo courtesy of Dresser Clark Division, Dresser Industries, Inc., Olean, NY.

polytropic processes are covered here—how they differ and why industry normally prefers the polytropic approach instead of the adiabatic approach for centrifugal compressor estimates.

Gas property calculations and the means by which mixture properties are determined are the focus of Chapter 3. Once reviewed, this chapter will serve as a reference for future calculations. The addition of water to dry gas mixtures and liquid knockout during compression interstage cooling are also discussed.

Chapter 4 looks at several ways of specifying inlet flow and examines the advantages and disadvantages of each. An understanding of the various inlet flow specifications presently used in industry is necessary for improved communications between engineers and between buyers and sellers of compression equipment.

Having developed the nomenclature and basic principles in the first four chapters, centrifugal compressor estimating procedures can now be discussed. Chapter 5 shows how to size centrifugal compressors of various configurations—straight through, intercooled, and side-load refrigeration. It shows how to determine the feasibility of satisfying process requirements; how to estimate the number of compressor bodies required for the compression process and the speed of compressor operation; and how to determine power requirements and estimate discharge temperatures. With a thorough understanding of these procedures, necessary process modifications can be made in the early stages of process design to optimize process-compressor compatibility. In turn, this will lead to improved overall plant efficiency and shorter lead-time to process development.

The centrifugal compressor performance curve is the subject of Chapters 6 and 7. Chapter 6 discusses the compressor characteristic and the importance of surge and stonewall. It reviews volume ratio effects—how the performance of one impeller affects the performance of those downstream—and discusses the various parameters that might be shown on a performance curve and indicates their general

A two-body, 114,000 ICFM, 30,000 HP, multi-stage centrifugal compressor train driven by a steam turbine. Photo courtesy of Dresser Clark Division, Dresser Industries, Inc., Olean, NY.

slopes. The fan laws, which can be used to estimate compressor speeds for other than design conditions, are also reviewed, along with system resistance and the relationship between the system resistance characteristics and the compressor performance curve.

Chapter 7 shows how to use compressor performance curves when operating conditions vary from manufacturers' curve conditions. It examines the effects of variation in inlet pressure, inlet temperature, molecular weight, *k*-value, and inlet compressibility and discusses the interaction of system resistance. Simple equations are developed to estimate the discharge pressure and horsepower resulting from inlet condition variations; the engineer will learn how to estimate the compressor speed changes necessary to overcome these variations. With this information, compressor off-design performance can be estimated without approaching the manufacturer, thus obtaining results more efficiently and in less time.

Chapter 8 discusses centrifugal compressor rerates—the reapplication of an existing compressor for increased capacity or for a different process altogether. A simple method is developed to test the feasibility of reusing an existing compressor rather than purchasing a new one. An understanding of this chapter will enable the engineer to evaluate the rerate potential of a compressor prior to approaching the manufacturer, allowing new process demands to be tailored to the limitations of the existing unit.

The appendixes include several tables and graphs for ease of compressor selection. In addition, they contain sufficient gas data—including gas properties, compressibility charts, Mollier charts, and atmospheric pressure charts—to enable the engineer to handle at least 95% of the applications for compressor estimates.

Ronald P. Lapina
June 1982

Nomenclature

a	speed of sound
b	number of compression sections
c_p	specific heat at constant pressure
c_v	specific heat at constant volume
d	impeller diameter
d_n	nozzle diameter
e	number of impellers on original rotor
f	number of blank stages on original rotor
g_c	universal gravitational constant
h	enthalpy
h_{2ad}	adiabatic, isentropic discharge enthalpy
H	head
H_{ad}	adiabatic head
H_p	polytropic head
k	ratio of specific heats
L_m	mechanical losses
m	mass flow
M	Mach number
MW	molecular weight
n	polytropic exponent
N	rotational speed
P	pressure
P_c	critical pressure
P_r	reduced pressure
P_w	partial pressure of water vapor

PWR	power	u_e	equivalent tip speed
PWR_g	gas power	v	specific volume
PWR_s	shaft power	V	gas velocity
Q	volume flow	V_a	actual nozzle velocity
r_p	pressure ratio	$V_{\max}$	maximum nozzle velocity
$r_{p\text{SECT}}$	sectional pressure ratio	W	work input
r_t	temperature ratio	x	mol fraction
r_v	volume ratio	y	mol fraction of water vapor
R	gas constant	Z	compressibility factor
SVF	standard volume flow	δ	flow coefficient
T	temperature	η_{ad}	adiabatic efficiency
T_c	critical temperature	η_p	polytropic efficiency
T_r	reduced temperature	θ	spead of sound index
u	mechanical tip speed	μ	head coefficient
		τ	torque

Subscripts

a	average	o	original
dg	dry gas	r	rated
fl	fan law	req	required
ic	inlet conditions	SL	side-load
L	liquid	std	standard
mix	mixture	w	water vapor
n	new	1*	inlet
nom	nominal	2*	discharge

One term that deserves consideration is ''stage.'' The compressor manufacturer normally uses ''stage'' to refer to one impeller with the associated stationary parts. It is common practice for the process or mechanical engineer to use ''stage'' to define the compression parts between successive flanges. In the author's opinion, ''stage'' is best used to define a single impeller and associated stationary parts, while ''section'' best defines the compression parts between successive flanges. In this text, these definitions will be used.

*In some instances, subscripts 1, 2, 3, and 4 refer to first section, second section, etc., as defined where they are used. Also, 1 and 2 may refer to first and second, as in gas constituents in Chapter 3.

1
The Centrifugal Compressor

The Multi-Stage Centrifugal Compressor

Centrifugal compressors find extensive use in the process industries today, since they are the most versatile and compact compression equipment in the flow range of 1,000–100,000 ICFM (1,700–170,000 m^3/h). The process industries use centrifugal compressors to elevate gas pressures for several purposes, some of which are:

1. To cause positive flow through a process. In this case the compressor elevates the pressure sufficiently to overcome process pressure drops due to piping, vessels, heat exchangers, valves, and fittings.
2. For separating heavy hydrocarbon mixtures. Once under pressure, the components of a hydrocarbon mixture can be separated by cooling the mixture in steps, the heavier hydrocarbons liquefying first.
3. To obtain supercool gases for refrigeration. Once under pressure, the gas temperature can be lowered by expansion of the gas.
4. As a means to force gas product flow through pipelines.

This chapter will look briefly at the basic components of the centrifugal compressor and the many arrangements available. Familiarization with the basic components and an awareness of the available compressor arrangements are necessary for an optimum compressor estimate.

This book describes how to estimate centrifugal compressor size and performance. However, the reader must understand that without the manufacturer's data, the results are only estimates. The results will provide a great

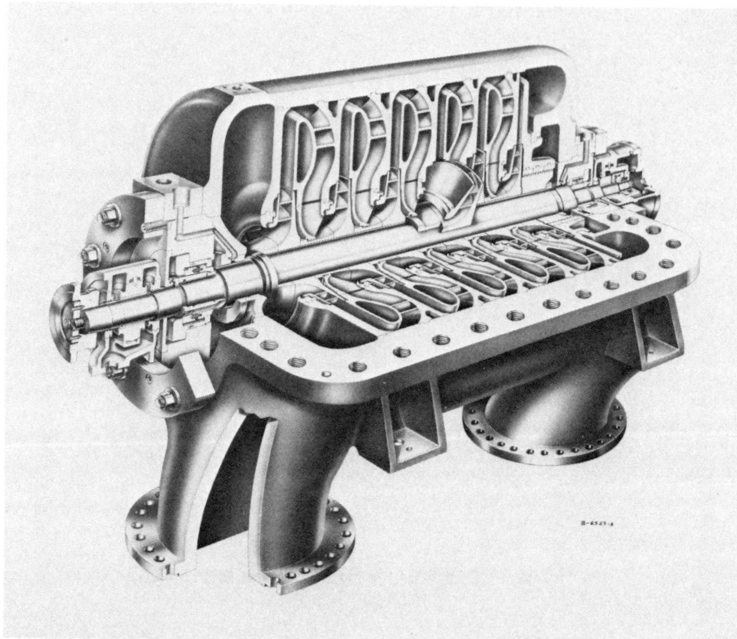

Figure 1-1a. Cutaway of a horizontally split centrifugal compressor. Photo courtesy of Ingersoll-Rand Company, Phillipsburg, NJ.

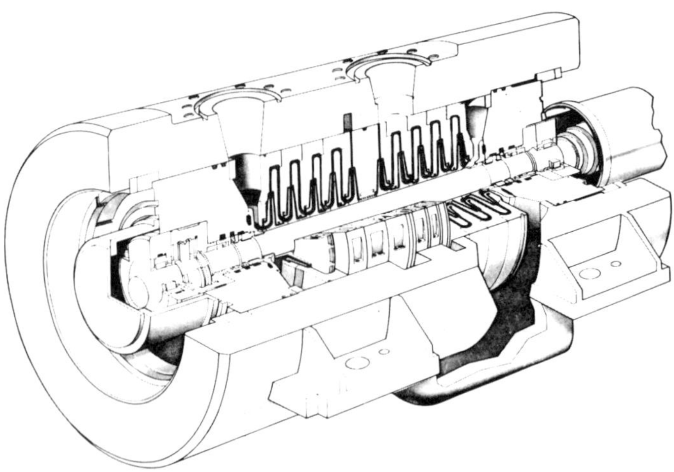

Figure 1-1b. Cutaway of a vertically split centrifugal compressor. Photo courtesy of Dresser Clark Division, Dresser Industries, Inc., Olean, NY.

deal of information during process design and in checking against manufacturers' selections and quoted performance. Still, the manufacturer must have some degree of freedom so that he may select the optimum and most efficient compressor design possible with his equipment.

Basic Components

The two main components of a multi-stage centrifugal compressor are the rotating element, called the rotor assembly, and the stationary parts, which include the casing and diaphragms. A cutaway view of a centrifugal compressor is shown in Figure 1-1.

The rotor assembly consists of the impellers (sometimes called wheels), sleeves, a balancing drum, and a thrust disc mounted on a shaft (see Figures 1-2, 1-3, and 1-4).

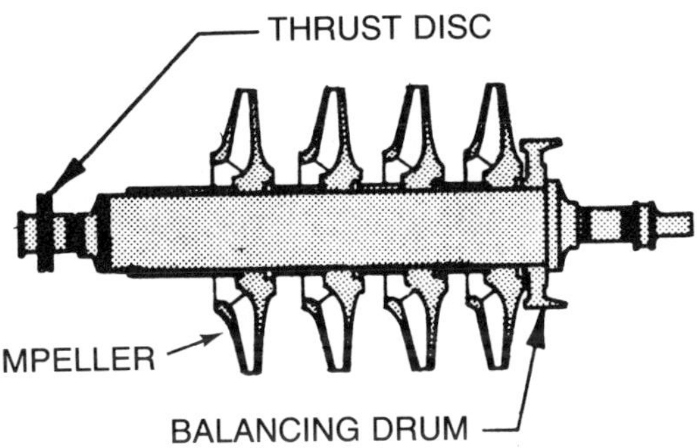

Figure 1-2. Components of the centrifugal compressor rotor assembly.

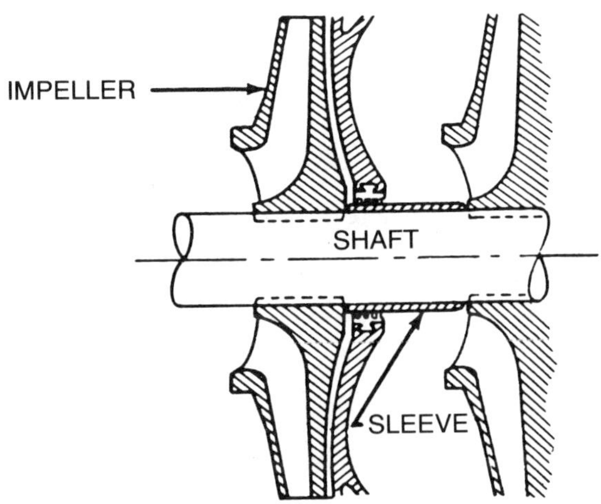

Figure 1-3. The impeller, shaft, and sleeve.

Figure 1-4. A centrifugal compressor rotor assembly set in the bottom half of a compressor casing. Photo courtesy of Elliott Company, Jeannette, PA.

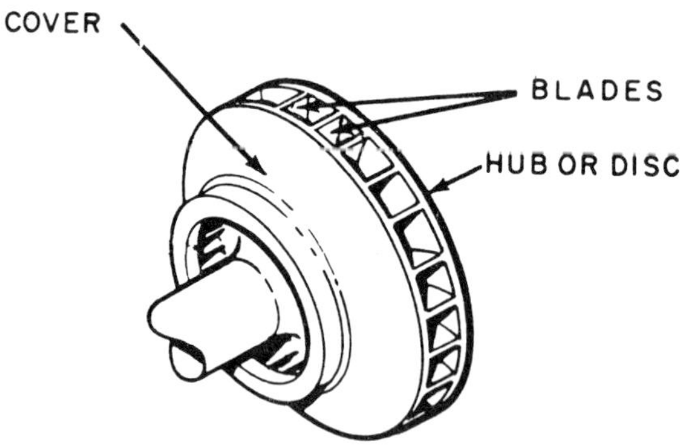

Figure 1-5. Components of the closed impeller.

Most multi-stage centrifugal compressors employ what is known as a closed impeller (see Figures 1-5 and 1-6). The closed impeller has three parts: the blades, a cover, and a hub or disc. The blades increase the velocity of the gas, thereby inducing a static pressure rise and imparting kinetic energy which converts to an additional pressure rise in the stationary parts of the compressor. The hub is that part of the impeller which attaches to the shaft. The hub confines the gas and acts to drive the blades. The cover also confines gas and, in addition, acts as a stiffener to prevent blade distortion from rotational forces.

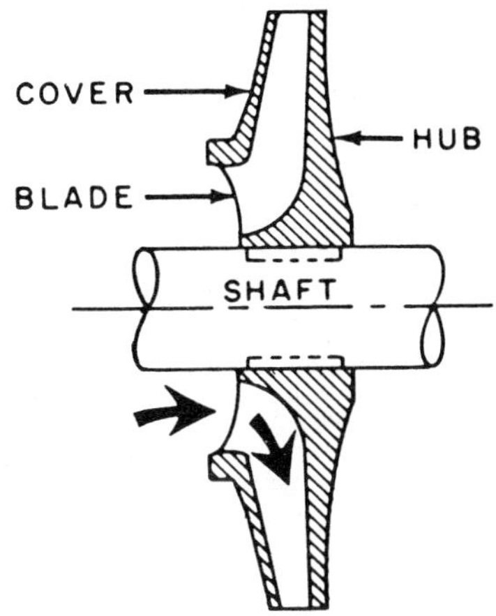

Figure 1-6. Cutaway of a closed impeller.

The manufacturer installs sleeves between adjacent impellers to position the impellers along the shaft. The sleeves act as wearing surfaces for inter-stage seals and also isolate the shaft from the process gas. The latter is especially beneficial when compressing corrosive gases. The manufacturer can make the sleeves, an inexpensive part, from a corrosion-resistant material and not worry about shaft corrosion. As a result, the manufacturer can make the shaft from a relatively inexpensive, high-strength material, giving the shaft high torque-carrying capabilities.

The balancing drum, or piston, is a thrust-equalizing component. It equalizes a large percentage of the rotor thrust. One end of the balancing drum is essentially at discharge pressure; the other end connects to a lower pressure point, normally the inlet end of the compressor, through a balancing-drum line. The resultant pressure differential across the balancing drum causes a thrust opposite to the impeller thrust. The amount of compensating thrust from the balancing drum is a function of the balancing-drum area. It is common practice to balance approximately 75% of the rotor thrust with the balancing drum.

The thrust disc fits between the thrust-bearing assemblies and, along with the thrust bearing, absorbs the net thrust of the rotor, i.e. that portion of the rotor thrust not equalized by the balancing drum. The thrust disc positions the rotor assembly in the casing such that the impellers line up with the gas passages in the stationary components.

The stationary components are formed by the casing, the diaphragms, and a special diaphragm called the inlet wall. These components form the various gas passages in the compressor: the inlet channel, the diffusers, the return bends, the return channels, and the discharge volute.

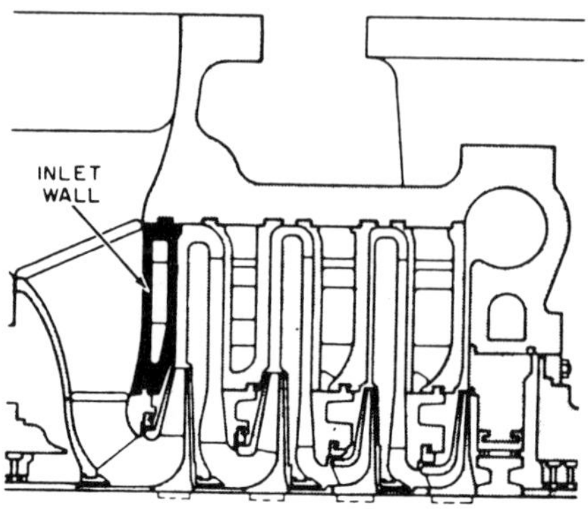

Figure 1-7. The inlet wall.

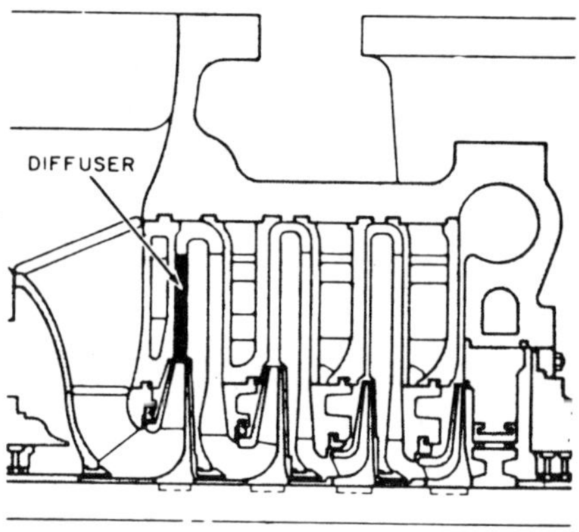

Figure 1-8. The diffuser.

The front face of the inlet wall (see Figure 1-7) forms part of the inlet channel of the compressor. The back face of the inlet wall is one of the two surfaces of the first diffuser (see Figure 1-8).

The diaphragm (Figures 1-9 and 1-10) is the stationary part which forms a face of the upstream diffuser, part of the return bend (Figure 1-11), all of the return channel (Figure 1-12), and a face of the downstream diffuser. Each impeller requires a diaphragm.

After the gas passes through the first impeller, it enters the diffuser, which systematically reduces the velocity of the gas by increasing the flow area radially. The kinetic energy imparted to the gas by the impeller is converted to pressure rise as a consequence of this velocity reduction in the diffuser.

After leaving the diffuser, the gas enters the return bend, which guides the gas into the return channel and finally into the next impeller. The return

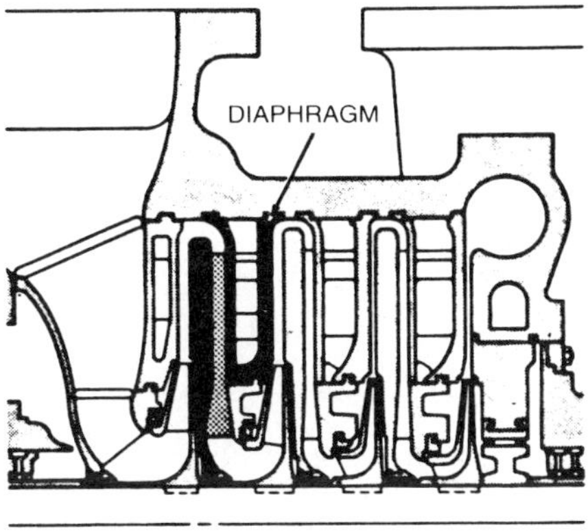

Figure 1-9. The diaphragm.

Figure 1-10. A diaphragm half being lowered in place into a centrifugal compressor casing. Photo courtesy of Elliott Company, Jeannette, PA.

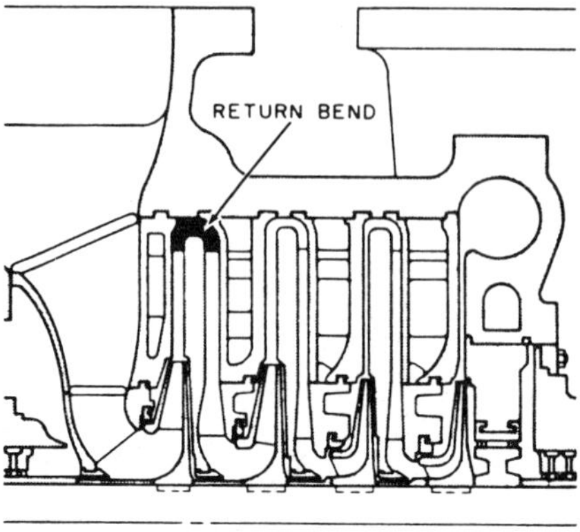

Figure 1-11. The return bend.

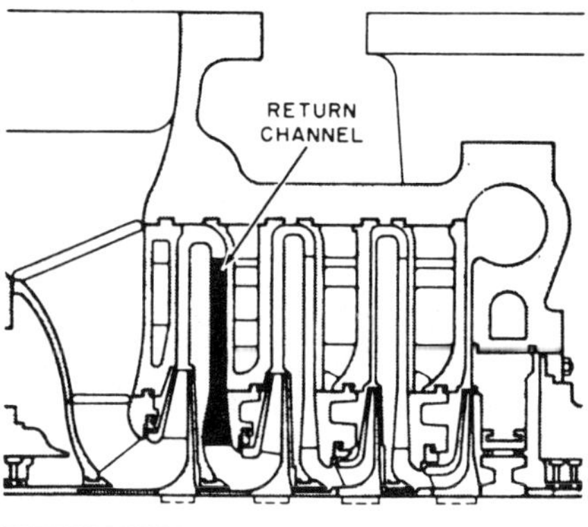

Figure 1-12. The return channel.

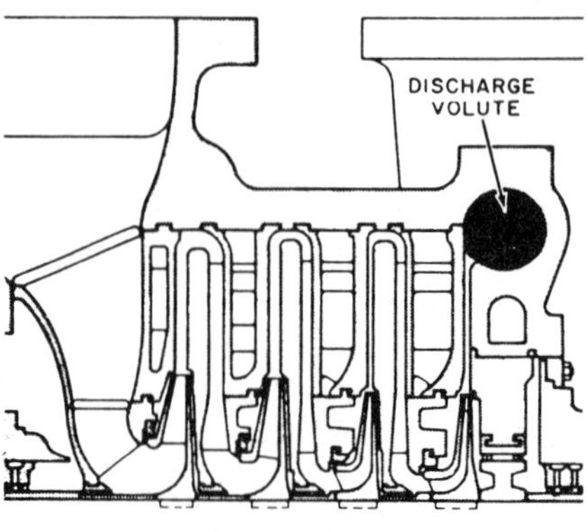

Figure 1-13. The discharge volute.

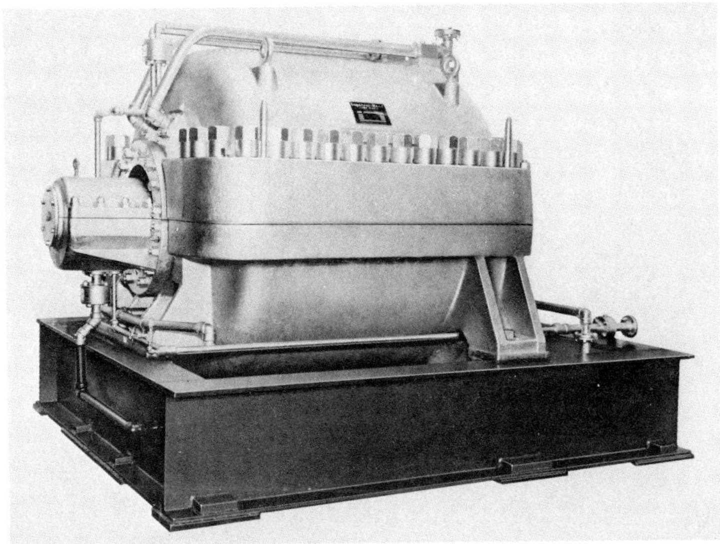

Figure 1-14. A horizontally split centrifugal compressor. Photo courtesy of Ingersoll-Rand Company, Phillipsburg, NJ.

channel includes guide vanes which eliminate swirl in the gas caused by rotation of the preceding impeller.

After the gas passes through the last diffuser, it enters the discharge volute (Figure 1-13), which directs the gas to the discharge nozzle and into the piping system.

The design of the entire gas passage is an important factor in the overall efficiency of the compression stage. The compression stage is defined here to include the return bend, return channel, impeller, and diffuser, respectively. The efficiency of the stage, then, is not only affected by the ability of the impeller to systematically increase the velocity of the gas, but also by the flow pattern entering the impeller and the uniformity of the area increase downstream.

Types of Casings

Centrifugal compressors are manufactured with two types of casings: the horizontally split casing (Figure 1-14) and the vertically split, or barrel, casing (Figure 1-15). Thc horizontally split casing is split along the rotor shaft and bolted at the split line. This type of casing generally employs either an up-nozzle arrangement, as illustrated in Figure 1-14, or a down-nozzle arrangement, although some intermediate nozzles may be located on the side to avoid nozzle interference. The down-nozzle horizontally split casing is preferable from a maintenance standpoint. If a horizontally split compressor

Figure 1-15. A vertically split centrifugal compressor. Photo courtesy of Ingersoll-Rand Company, Phillipsburg, NJ.

Figure 1-16. Vertically split centrifugal compressor showing method of removal of internals. Photo courtesy of Dresser Clark Division, Dresser Industries Incorporated, Olean, NY.

is ordered with down nozzles, the compressor casing can be opened and the rotor assembly lifted with an overhead crane without disturbing process piping. Space requirements for maintenance are therefore minimal.

At about 500 psig (3,500 kPa), horizontally split casings lose their sealing capability, resulting in an increased potential for the escape of process gas. Above this level, manufacturers will resort to the vertically split, or barrel, compressor. (The actual pressure split depends on manufacturer and the compressor size. The pressure split will be lower for large compressors and higher for small compressors.) This type of casing has the rotor assembly housed in a complete cylinder. Rotor removal occurs at the ends of the compressor by sliding the rotor assembly (and diaphragms or inner casing) out of the barrel (see Figure 1-16). As a result, sufficient space must be available for removing and maintaining the rotor assembly.

Figures 1-17, 1-18, and 1-19 show multi-body centrifugal compressor trains in various stages of development, i.e. assembly, testing, and installation, respectively.

Arrangements

There are several different arrangements for centrifugal compressors. The basic arrangement is the straight-through compressor (Figure 1-20). This configuration has only two nozzles, an inlet and a discharge. Gas enters the

Figure 1-17. A multi-body centrifugal compressor train undergoing assembly. Photo courtesy of Ingersoll-Rand Company, Phillipsburg, NJ.

Figure 1-18. A large turbine-driven, multi-body centrifugal compressor train undergoing testing. Photo courtesy of Elliott Company, Jeannette, PA.

Figure 1-19. A large turbine-driven, multi-body centrifugal compressor train installed at job site. Photo courtesy of Dresser Clark Division, Dresser Industries, Inc., Olean, NY.

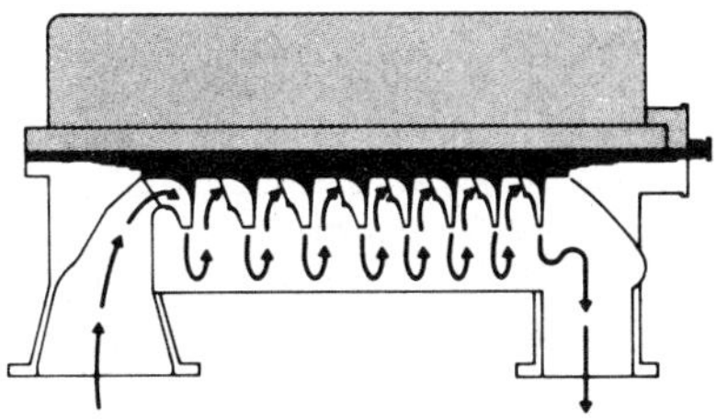

Figure 1-20. Schematic of a straight-through centrifugal compressor.

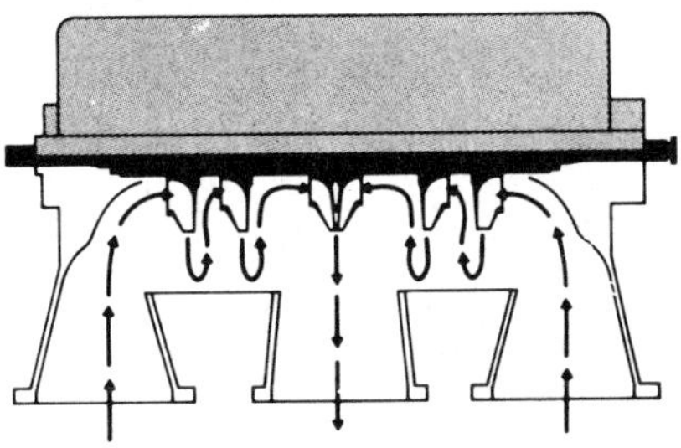

Figure 1-21. Schematic of a double-flow centrifugal compressor.

inlet nozzle, passes through one or more stages of compression, and leaves through the discharge nozzle into the piping system.

When flow requirements are large and head requirements low, the manufacturer can often reduce the compressor size by employing the double-flow arrangement (Figure 1-21). In this configuration the inlet flow is split into two equal portions, each entering an end of the compressor. The gas proceeds through as many as four compression stages and collects in the discharge volute. The gas then exits through a common discharge nozzle at the center of the compressor. This configuration saves cost, as the outside diameter of the casing is smaller than would be necessary with the single-flow arragement. In effect, this approach doubles the capacity of the compressor casing. However, by using this type of casing, the manufacturer sacrifices potential work output from the compressor, as this arrangement can only produce about half the work output of the straight-through arrangement. The reasons for this will become more evident in Chapter 5.

For situations where gas temperatures due to compression become too high, one can maintain reasonable discharge temperatures and actually improve compression efficiency by cooling the gas part way through the compression process. One way to accomplish this is to use two compressor bodies (casings) with an external heat exchanger in-between. However, compressor manufacturers can furnish an intercooled compressor (Figure 1-22). This arrangement is actually two compressors in one casing. Gas enters

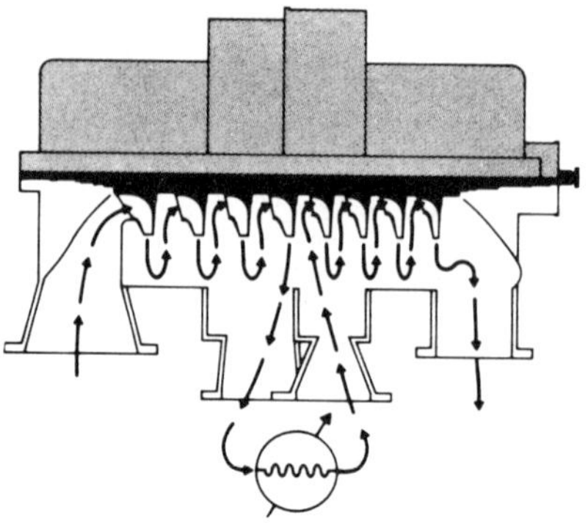

Figure 1-22. Schematic of a single intercooled centrifugal compressor.

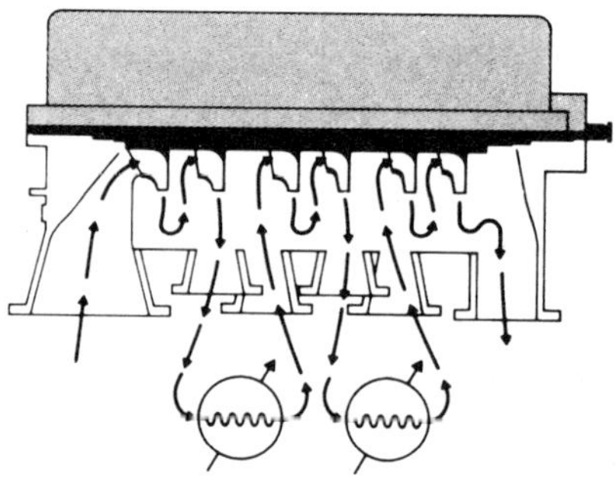

Figure 1-23. Schematic of a double intercooled centrifugal compressor.

the compressor through the inlet flange, proceeds through several stages of compression, and exits through an intermediate discharge nozzle to an external heat exchanger. After being cooled, the piping system redirects the gas into the middle of the compressor casing, where it goes through several more stages of compression. The gas then proceeds to the final discharge nozzle and into the piping system.

High-molecular-weight gases heat more rapidly during compression than low-molecular-weight gases. As a result, additional cooling points are sometimes necessary for heavy gases. For these cases the manufacturer can often employ the double intercooled compressor (Figure 1-23) to keep the compression process in a single casing. (Depending on the head requirement from the compressor, this may not be possible.) In this arrangement, gas leaves the compressor at two intermediate locations for cooling.

Each external cooler requires two additional nozzles on the casing, and the diaphragm for the cooler requires more axial space than a normal diaphragm. As a result, arranging a compressor for intercooling increases the

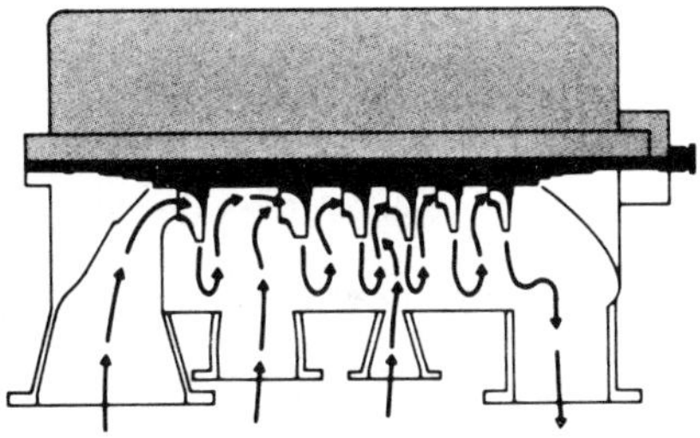

Figure 1-24. Schematic of a side-load centrifugal compressor.

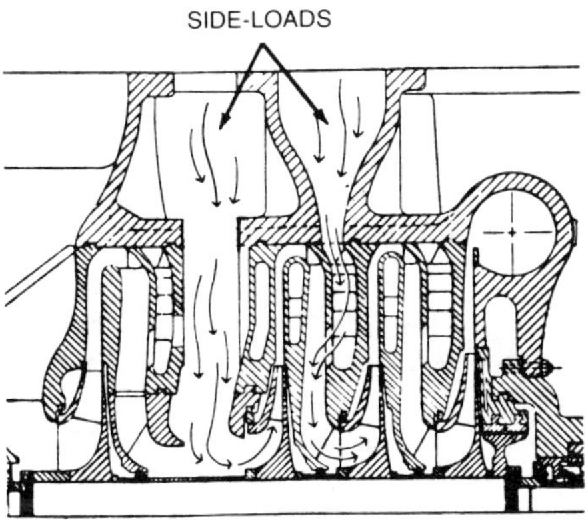

Figure 1-25. Two approaches for injecting side-loads: the blank stage (left) and the diaphragm core (right).

rotor length between bearing centers. Rotor critical speeds limit the maximum allowable distance between bearing centers, so with limited bearing span, nozzle interference or impeller axial space requirements could become a problem if too many cool points are attempted. Therefore, two cool points are the maximum feasible number for a single compressor casing.

Some processes, such as refrigeration processes, require several pressure levels from the compressor. The manufacturer can accomplish this by using the side-load or side-stream compressor (Figure 1-24). In this arrangement, the main gas stream enters the compressor inlet nozzle. At the required pressure level, another gas stream enters through an intermediate inlet nozzle. The two streams mix inside the casing and enter the next compression stage. Manufacturers inject the side-loads in two ways (see Figure 1-25). When side-load flows are small, the side-load may be injected through the core of a diaphragm. If the side-load flow is large, say more than 10% of the previous section flow, the side-load may be injected through a blank-stage space. The latter approach is more efficient, as the manufacturer can reduce

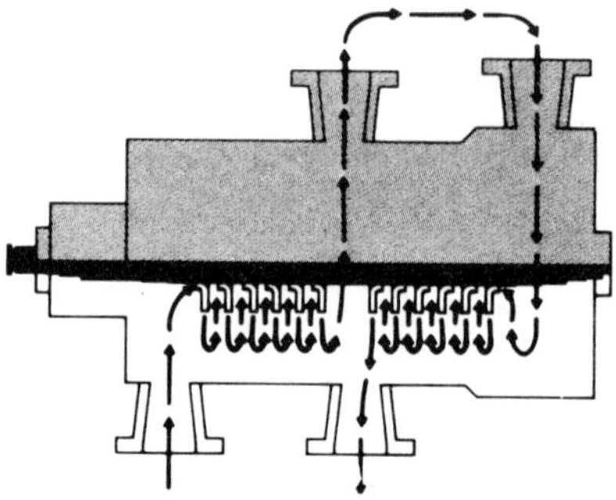

Figure 1-26. Schematic of a back-to-back centrifugal compressor.

pressure drop and better control the mixing of the two streams (note the contoured flow pattern and generous area for side-load entry and mixing in Figure 1-25). The former is less expensive and, more important, does not lengthen the rotor as does the latter approach.

A single compressor body can normally incorporate up to three side-loads. Only nozzle interference and the axial space requirements of the impellers and mixing area limit the feasible number of side-loads. As in the case of the intercooled compressor, the main limiting consideration is the rotor critical speeds.

An intermediate extraction of process gas is also possible with this configuration. In this case, some of the gas flows through the intermediate extraction nozzle and into the piping system. The remaining gas flows to the following compression stages.

It may be interesting to note that some cooling takes place in a side-load compressor. This phenomenon occurs because the stream injected at an intermediate nozzle is normally at or slightly above the saturation temperature and therefore cooler than the main stream, which was heated by compression. Mixture of the two gases will lower the temperature of the main stream.

In high-pressure applications, compensation for rotor thrust becomes extremely difficult, especially during process upset. The manufacturer can overcome the high rotor thrust by using the back-to-back arrangement (Figure 1-26). In this configuration the process gas enters the suction flange, proceeds through several stages of compression, and exits the casing through an intermediate discharge nozzle. The gas then reenters the far end of the compressor through a second inlet nozzle. (This crossover of gas flow can occur either internal or external to the compressor casing.) After more compression, the gas flows to the final discharge nozzle, which is located in the center of the compressor. With this arrangement, approximately half the impellers face one direction and the rest face the opposite direction. The thrust from the two rotor halves oppose each other, resulting in partial thrust

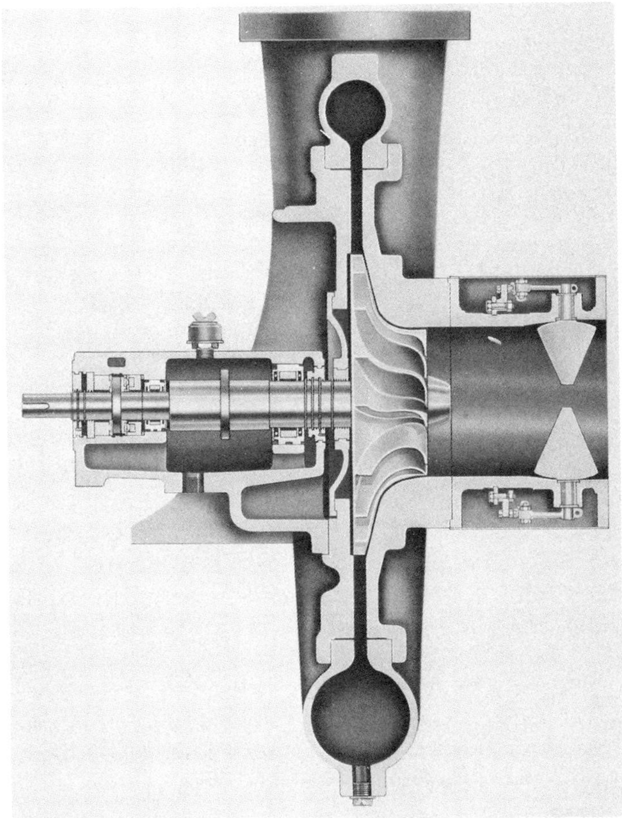

Figure 1-27. Cutaway of the single-stage centrifugal compressor. Photo courtesy of Roots Operations, Dresser Industries, Inc., Connersville, IN.

compensation. When necessary, due to high compression temperatures, the process gas can be directed through an external heat exchanger between the intermediate discharge nozzle and the second inlet nozzle.

The Single-Stage Centrifugal Compressor

For cases where the pressure ratio, and therefore the head requirement, is low, the single-stage centrifugal compressor is employed (see Figures 1-27 and 1-28). Single-stage compressors are normally used for head requirements up to about 15,000 ft-lbf/1bm (45 k·Nm/kg), and some manufacturers have capabilities up to 30,000 ft-lbf/lbm (90 k·Nm/kg). To gain an appreciation for the use of the single-stage compressor, 15,000 ft-lbf/lbm

Figure 1-28. A motor-driven single-stage centrifugal compressor. Photo courtesy of Roots Operations, Dresser Industries, Inc., Connersville, IN.

of adiabatic head equates to a pressure ratio of approximately 1.6 for air applications. The most extensive use of single-stage centrifugal compressors is in low-pressure air service. However, they find considerable application as low-pressure process compressors. They are also used in high-pressure process applications where required pressure ratios are within their capabilities.

The single-stage centrifugal compressor looks and operates very much like a single-stage centrifugal process pump. The impeller is overhung, i.e. located outboard of the two radial bearings opposite the end of the driver. (This is very different from the multi-stage compressor design. Reference to the multi-stage compressor, Figures 1-1 and 1-4, will show that all impellers are located between the bearings.) The diffuser and volute sections, which slow the gas and convert kinetic energy to pressure, are essentially part of the casing.

Gas enters the compressor axially (parallel to the axis of the shaft) and proceeds through the impeller. The impeller induces a static pressure rise and imparts kinetic energy to the gas. After leaving the impeller, the gas proceeds to the diffuser and volute sections, which systematically slow the gas and convert the kinetic energy to additional pressure rise. The gas then exits the discharge nozzle, which can be oriented in several angles to match process piping considerations.

2
Thermodynamics

Ideal Gas Laws

The thermodynamic law fundamental to an understanding of compressor operation is the ideal gas law, which is expressed in equation form as follows.

$$Pv = RT \tag{2-1}$$

where:

P = pressure
v = specific volume
T = absolute temperature
R = gas constant, and is a function of the molecular weight of the gas. (R = 1,545/MW ft-lbf/lbm-°R in the English system and R = 8,314/MW N·m/kg·K in the metric system. MW = molecular weight.)

This equation shows that if the volume of a gas is held constant and the gas heated, the pressure will increase proportionally to the absolute temperature increase. If the temperature of the gas is held constant, the change in pressure will be inversely proportional to the change in volume. Furthermore, R is a function only of the molecular weight of the gas and is constant for any given pure gas or gas mixture. Therefore, any changes in pressure, volume, or temperature must take place such that the following relationship holds:

$$\frac{Pv}{T} = \text{constant}$$

No gas conforms exactly to this law; however, the direction of change of any parameter resulting from a change in any other parameter will be as the ideal gas law implies. Note that all gases approach the ideal gas either at very low pressures or at very high temperatures.

The amount of deviation from the ideal gas law is a function of the gas and the pressure and temperature at which it operates. The ideal gas law can be made a real gas law by introducing a deviance parameter defined as the compressibility factor Z. Thus, the ideal gas equation is modified as follows:

$$Pv = ZRT \tag{2-2}$$

This equation is an exact relationship for all real gases, since there is always a relationship between the pressure, volume, and temperature of the gas. In order to make this equation useful, however, it was necessary to develop a universal method of calculating the compressibility factor. The generalized compressibility charts, which can be found in numerous places in the literature and in Appendix C, are the result of this study. The compressibility charts are formulated as a function of the reduced pressure (P_r) and reduced temperature (T_r) of the gas or gas mixture. The reduced pressure is derived by dividing the actual pressure of the gas by the critical pressure:

$$P_r = \frac{P}{P_c} \tag{2-3}$$

and similarly:

$$T_r = \frac{T}{T_c} \tag{2-4}$$

From these two equations, one can see the primary assumption upon which the compressibility charts are based. The assumption is known as the Law of Corresponding States, which expresses that any two gases that are at the same percentage of their respective critical pressure and critical temperature behave alike. This is not exactly true, thereby making the value of the compressibility factor obtained from the charts an approximation. This approximation is very close to actual behavior for most gases, however, making the approach very useful for engineering calculations.

Equation 2-2 is known as an equation of state. There are several equations of state available in the literature that are more accurate, but none are as easy to manipulate or as convenient to use as Equation 2-2. This equation forms the basis for the most widely used approach for performing compressor estimates.

Compressor requirements can be more accurately determined from a Mollier chart. Such charts are available; however, they are not readily available for all pure gases or for gas mixtures. For these cases the reader will have to resort to Equation 2-2. (Note that one can construct a Mollier chart for any gas or gas mixture; however, the reader will not normally have access to such a chart when performing a compressor estimate on gases other than those whose charts are readily available in the literature or in Appendix C.)

Adiabatic and Polytropic Processes

An adiabatic process is defined as a process in which no heat transfer takes place. This does not mean that the temperature is constant, but rather that there is no heat transferred into or out of the system. Compressor manufacturers normally use the term "adiabatic" to mean "adiabatic isentropic (constant entropy)." Thus, when speaking of adiabatic head, one normally means adiabatic, isentropic head. The adiabatic, isentropic process is of the form:

$$Pv^k = \text{constant}$$

The polytropic process is a variable entropy process in which heat transfer can take place. It follows Equation 2-5. Either of these processes would be reversible if the efficiency were 100%. Note that the adiabatic process is a special case of the more general polytropic process:

$$Pv^n = \text{constant} \tag{2-5}$$

where:

n = k(ratio of specific heats c_p/c_v) for an adiabatic, isentropic process
n = 1 for an isothermal process (constant temperature)
n = 0 for an isobaric process (constant pressure)
n = ∞ for an isometric process (constant volume)

All compressor applications are of the type defined by Equation 2-5.

By manipulating Equation 2-5, one can derive the pressure ratio (r_p), volume ratio (r_v), and temperature ratio (r_t) equations, which find extensive use in compressor calculations.

Rewriting Equation 2-5 yields the following:

$$P_1v_1^n = P_2v_2^n = \text{constant}$$

$$\frac{P_2}{P_1} = \left(\frac{v_1}{v_2}\right)^n$$

Now define:

$$r_p = \frac{P_2}{P_1}$$

and:

$$r_v = \frac{v_1}{v_2}$$

Then:

$$r_p = r_v^n \tag{2-6}$$

and:

$$r_v = r_p^{1/n} \tag{2-7}$$

Next introduce the ideal gas equation to obtain a relationship for the temperature ratio:

$$\frac{P_2}{P_1} = \left(\frac{v_1}{v_2}\right)^n$$

$$Pv = RT$$

$$\frac{P_1 v_1}{T_1} = \frac{P_2 v_2}{T_2} = R = \text{constant}$$

$$\frac{v_1}{v_2} = \frac{P_2}{P_1}\frac{T_1}{T_2}$$

$$\frac{P_2}{P_1} = \left(\frac{P_2}{P_1}\frac{T_1}{T_2}\right)^n$$

$$\frac{P_2}{P_1} = \left(\frac{P_2}{P_1}\right)^n \left(\frac{T_1}{T_2}\right)^n$$

$$\left(\frac{P_2}{P_1}\right)^{1-n} = \left(\frac{T_1}{T_2}\right)^{n} = \left(\frac{T_2}{T_1}\right)^{-n}$$

$$\frac{P_2}{P_1} = \left(\frac{T_2}{T_{11}}\right)^{n/(n-1)}$$

$$r_p = r_t^{n/(n-1)} \tag{2-8}$$

and:

$$r_t = r_p^{(n-1)/n} \tag{2-9}$$

Similarly, a relationship between the temperature ratio and the volume ratio can be obtained:

$$r_v = r_t^{1/(n-1)} \tag{2-10}$$

$$r_t = r_v^{n-1} \tag{2-11}$$

The derivations of Equations 2-10 and 2-11 are left to the reader.

Note that if k is substituted for n in Equations 2-6 through 2-11, the results are the pressure ratio, volume ratio, and temperature ratio relationships for an adiabatic process providing the adiabatic efficiency were 100% and therefore reversible. All of the equations are presented here for convenience:

Polytropic Process	**Adiabatic Process (when the adiabatic efficiency = 100%)**
$Pv^n = constant$	$Pv^k = constant$
$r_p = r_v^n$	$r_p = r_v^k$
$r_v = r_p^{1/n}$	$r_v = r_p^{1/k}$
$r_p = r_t^{n/(n-1)}$	$r_p = r_t^{k/(k-1)}$
$r_t = r_p^{(n-1)/n}$	$r_t = r_p^{(k-1)/k}$
$r_v = r_t^{1(n-1)}$	$r_v = r_t^{1(k-1)}$
$r_t = r_v^{n-1}$	$r_t = r_v^{k-1}$

These relationships for the polytropic process already have the polytropic efficiency included in the n-value and therefore find extensive use in compressor calculations as tabulated. The volume ratio thus calculated differs from the actual volume ratio only by the ratio of the inlet and discharge compressibility factors. The adiabatic relationships, on the other hand, derive from an adiabatic, isentropic process. Therefore, *the preceding adiabatic relationships only apply when the adiabatic compression efficiency is 100%*. The actual discharge temperature will be much higher than that calculated by the adiabatic relationships. A discharge-temperature equation for adiabatic compression processes with $\eta_{ad} < 100\%$ follows later in this chapter.

Head, Work, and Efficiency

The general form of the thermodynamic head equation for a polytropic process is:

$$H_p = ZRT_1 \frac{n}{n-1} [r_p^{(n-1)/n} - 1] \tag{2-12}$$

This equation derives from integrating the steady-state, steady-flow work equation given by:

$$H_p = \int vdP$$

Refer to Appendix A for the derivation.

Note that R, the gas constant, is inversely proportional to the molecular weight:

$$R = \frac{\text{constant}}{MW}$$

Therefore, the amount of head required to compress a gas is inversely proportional to the molecular weight. Note also that the head requirement is directly proportional to the inlet temperature and required pressure ratio. The required head is directly proportional to the compressibility factor. This means that the lower the compressibility factor, the lower the head requirement and therefore the lower the work input. An assumed value of $Z = 1.0$ will nearly always be conservative. Since Z values can go to .75 and lower during compression, however, assuming that $Z = 1.0$ may be so conservative as to be useless.

The adiabatic head equation derives from Equation 2-12 by substituting k for n:

$$H_{ad} = ZRT_1 \frac{k}{k-1} [r_p^{(k-1)/k} - 1]$$

The value of n is a function of the k-value of the gas and the polytropic efficiency. The relationship is shown by Equation 2-13:

$$\frac{n}{n-1} = \frac{k}{k-1}\eta_p \tag{2-13}$$

where:
η_p = polytropic efficiency

The derivation of Equation 2-13 is in Appendix A.

The efficiency of a thermodynamic process is the ratio of the work output to the work input. The work output per mass of gas compressed is the polytropic head. The work input per mass of gas compressed is defined by:

$$W = \frac{H_p}{\eta_p}$$

When combined with the mass flow and the proper conversion constants, the work becomes the gas power requirement.

$$PWR = \frac{mH_p}{C_1\eta_p} \tag{2-14}$$

where:
m = mass flow
C_1 = conversion constant

Likewise, for adiabatic compression:

$$W = \frac{H_{ad}}{\eta_{ad}}$$

and the power requirement becomes:

$$PWR = \frac{mH_{ad}}{C_1\eta_{ad}}$$

When the compressor is installed in the field, the power required from the driver will be the same whether the process is considered adiabatic or polytropic during compressor design. Therefore, the work input must be the same value whether the process is considered adiabatic or polytropic. Thus:

$$W = \frac{H_{ad}}{\eta_{ad}} = \frac{H_p}{\eta_p}$$

The only precaution that must be taken is that corresponding values be used: *When using adiabatic head, use adiabatic efficiency; when using polytropic head, use polytropic efficiency.*

Why Use the Polytropic Process?

In the preceding discussion on the work equation, it was shown that the correct power requirement would evolve whether the process was considered adiabatic or polytropic. Why, then, should engineers doing compressor estimates consider the process polytropic? After all, it would appear that the adiabatic approach is simpler and quicker, since it does not require calculating an n-value.

Actually, the polytropic calculation is easier, especially in the determination of the discharge temperature (this will be covered later).

The polytropic approach offers two distinct advantages over the adiabatic approach: First is the advantage of the polytropic efficiency. Equation 2-13 expresses the relationship between the n-value, k-value, and the polytropic efficiency. This relationship follows in a slightly different form for convenience:

$$\eta_p = \frac{\dfrac{n}{n-1}}{\dfrac{k}{k-1}}$$

From this relationship it can be deduced that the polytropic efficiency is independent of the thermodynamic state of the gas undergoing compression.

The equation for the adiabatic efficiency is derived as follows:

$$W = \frac{H_{ad}}{\eta_{ad}} = \frac{H_p}{\eta_p}$$

$$\frac{H_{ad}}{\eta_{ad}} = \frac{H_p}{\eta_p}$$

$$\eta_{ad} = \frac{H_{ad}}{H_p}\eta_p$$

$$= \frac{ZRT_1 \dfrac{k}{k-1}[r_p^{(k-1)/k} - 1]}{ZRT_1 \dfrac{n}{n-1}[r_p^{(n-1)/n} - 1]}\eta_p$$

$$= \frac{ZRT_1 \dfrac{k}{k-1} [r_p^{(k-1)/k} - 1]}{ZRT_1 \dfrac{k}{k-1} \eta_p [r_p^{(n-1)/n} - 1]} \eta_p$$

$$= \frac{r_p^{(k-1)/k} - 1}{r_p^{(n-1)/n} - 1}$$

As can be seen, the adiabatic efficiency is a function of the pressure ratio and therefore dependent upon the thermodynamic state of the gas undergoing compression.

As an example, refer to Figure 2-1, which is a graph of adiabatic efficiency as a function of pressure ratio for a gas with a *k*-value of 1.4 and an impeller with a polytropic efficiency of 78%.

if $r_p = 2.0, \quad \eta_{ad} = 75.8\%$

if $r_p = 10.0, \quad \eta_{ad} = 70.3\%$

The two values for adiabatic efficiency are quite different.

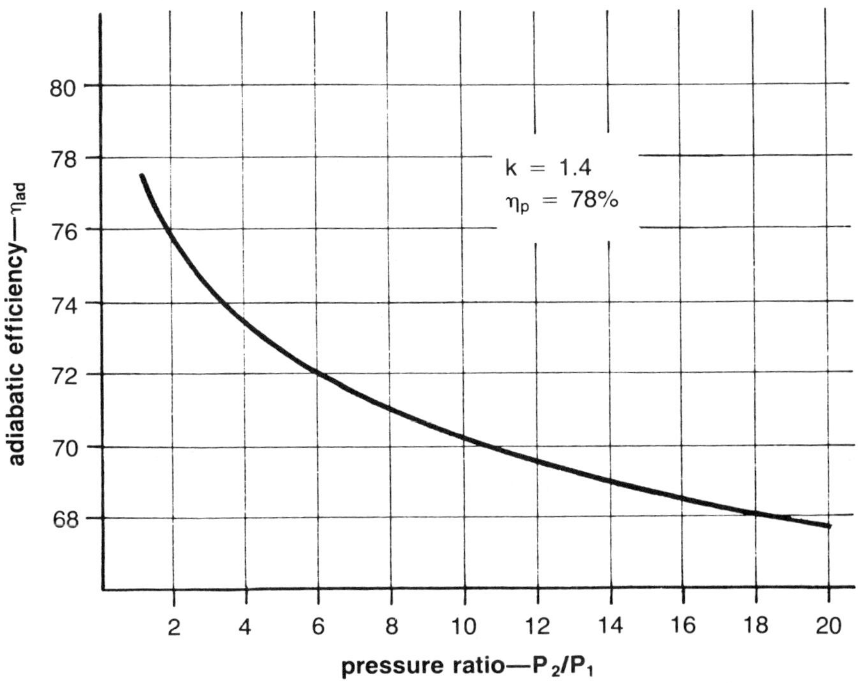

Figure 2-1. Adiabatic efficiency as a function of pressure ratio for k = 1.4 and polytropic efficiency = 78%.

Thus if a manufacturer considers all processes polytropic, he can design an impeller, determine its efficiency curve, and apply it without correction regardless of the pressure, temperature, or molecular weight of the gas undergoing compression. He does not have this advantage if he considers the process adiabatic. In fact, he would have to construct a whole series of efficiency charts for various pressure ratios and interpolate between them for each application. Likewise, if the reader has data indicating the magnitude of the polytropic efficiency of a centrifugal compressor, he can apply the data without modification for pressure ratio.

Note that for low-pressure-ratio applications, such as low-pressure air service, the effects of pressure ratio on the adiabatic efficiency are negligible and the adiabatic approach works well. For these applications, the reader should perform his estimates and communicate in terms of the adiabatic head and adiabatic efficiency.

The other advantage to the polytropic approach concerns the head: The sum of the polytropic heads for each stage of compression equals the total polytropic head required to get from state-point 1 to state-point 2 (see Figure 2-2). This is not true for adiabatic heads. This is better seen through the following example:

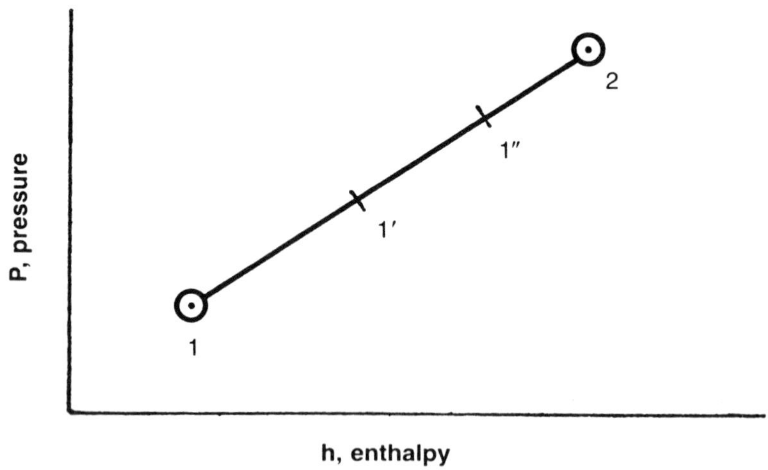

$$H_{p\,1-2} = H_{p\,1-1'} + H_{p\,1'-1''} + H_{p\,1''-2}$$

$$H_{ad\,1-2} \neq H_{ad\,1-1'} + H_{ad\,1'-1''} + H_{ad\,1''-2}$$

Figure 2-2. The sum of the stage polytropic heads equals the total polytropic head of the compressor.

EXAMPLE 2-1

Assume a four-stage compressor. The required overall pressure ratio of the process is 2.9. The gas has the following properties:

$$\begin{aligned} k &= 1.4 \\ MW &= 28 \text{ lbm/lbm-mol (kg/kg·mol)} \\ Z_a &= 1.0 \\ T_1 &= 100°\text{F } (37.8°\text{C}) = 560°\text{R } (311 \text{ K}) \end{aligned}$$

For this example, assume that the polytropic efficiency of each stage is 78%. Use Figure 2-1 to determine the adiabatic efficiency both overall and for each stage of compression.

First rewrite the polytropic head equation with the pertinent conversion factors. The adiabatic head equation will follow by substituting k for n. Note at this point that the head equation should be based on the average compressibility, Z_a. (This fact will be explored in greater detail in Chapter 5, "Why Use Average k- and Z-values?")

English units:

$$H_p = Z_a \frac{1{,}545}{MW} T_1 \frac{n}{n-1} [r_p^{(n-1)/n} - 1] \tag{2-15}$$

metric units:

$$H_p = \frac{Z_a}{1{,}000} \frac{8{,}314}{MW} T_1 \frac{n}{n-1} [r_p^{(n-1)/n} - 1] \tag{2-16}$$

Next calculate the value of $n/(n-1)$ from the k-value and the polytropic efficiency:

$$\frac{k}{k-1} = \frac{1.4}{0.4} = 3.5$$

$$\frac{n}{n-1} = \frac{k}{k-1} \eta_p = (3.5)(0.78) = 2.73$$

Now calculate the overall polytropic and adiabatic heads for both the English and metric systems:

English:

$$H_p = (1.0)\left(\frac{1{,}545}{28}\right)(560)(2.73)[(2.9)^{1/2.73} - 1]$$

$$= 40{,}240 \frac{\text{ft-lbf}}{\text{lbm}}$$

$$H_{ad} = (1.0)\left(\frac{1{,}545}{28}\right)(560)(3.5)[(2.9)^{1/3.5} - 1]$$

$$= 38{,}450 \, \frac{\text{ft-lbf}}{\text{lbm}}$$

metric:

$$H_p = \left(\frac{1}{1{,}000}\right)\left(\frac{8{,}314}{28}\right)(311)(2.73)[(2.9)^{1/2.73} - 1]$$

$$= 120.2 \text{ kN·m/kg}$$

$$H_{ad} = \left(\frac{1}{1{,}000}\right)\left(\frac{8{,}314}{28}\right)(311)(3.5)[(2.9)^{1/3.5} - 1]$$

$$= 114.9 \text{ kN·m/kg}$$

Also note from Figure 2-1 that the overall adiabatic efficiency for a pressure ratio of 2.9 is 74.5%.

For the purpose of this discussion, assume that each stage develops an equivalent pressure ratio. The pressure ratio per stage of a four-stage compressor would then be:

$$r_p/\text{stage} = 2.9^{1/4} = 1.305$$

The discharge temperature equations for the polytropic and adiabatic processes are as follows:

Polytropic	**Adiabatic**
$r_t = r_p^{(n-1)/n}$ $T_2 = T_1\, r_p^{(n-1)/n}$	$T_2 = \dfrac{T_1(r_p^{(k-1)/k} - 1)}{\eta_{ad}} + T_1$

Note that the proof of the adiabatic discharge-temperature equation will be evident from this example. The discharge temperature must be the same regardless whether one considers the process adiabatic or polytropic. Reference to Table 2-1 will show that in fact:

$$T_2 = T_1 r_p^{(n-1)/n} = \frac{T_1(r_p^{(k-1)/k} - 1)}{\eta_{ad}} + T_1$$

Now determine the polytropic and adiabatic heads required for each stage of compression and sum them. The adiabatic efficiency for each stage can be derived from Figure 2-1. Table 2-1 gives the English units; the metric units are given in parentheses. This approach will be adopted where applicable throughout this text.

As can be seen in Table 2-1, the sum of the individual polytropic heads of each stage add up to the total polytropic head required for the overall pressure ratio. One can also see from this table that this principle does not hold true for adiabatic heads.

Table 2-1
Stage-by-Stage Calculation

Stage	r_p	η_p (%)	η_{ad} (%)	T_1 °R (K)	H_p ft-lbf/lbm (kN·m/kg)	H_{ad} ft-lbf/lbm (kN·m/kg)	T_{2p} °R (K)	T_{2ad} °R (K)
1	1.305	78	77.2	560 (311)	8,640 (25.8)	8,550 (25.5)	617 (343)	617 (343)
2	1.305	78	77.2	617 (343)	9,520 (28.5)	9,420 (28.2)	681 (378)	681 (378)
3	1.305	78	77.2	681 (378)	10,510 (31.4)	10,390 (31.0)	750 (417)	750 (417)
4	1.305	78	77.2	750 (417)	11,570 (34.6)	11,450 (34.2)	827 (459)	827 (459)
Results based on stage-by-stage	2.9	—	—	—	40,240 (120.3)*	39,810 (118.9)	—	—
Results based on overall pressure ratio	2.9	78	74.5	—	40,240 (120.2)	38,450 (114.9)	827 (459)	827 (459)

*Difference due to round-off error.

In actual compression processes, the head output per stage tends to be nearly equal, while the pressure ratio decreases due to the increasing inlet temperature as the gas proceeds from one stage to another. The preceding example held the pressure ratio across each stage constant for simplicity. As an exercise, the reader may wish to hold the polytropic head per stage constant at, say, 10,000 ft-lbf/lbm (30 N·m/kg) and calculate the pressure ratio. Once the pressure ratios are known, the adiabatic heads can be calculated and the principle discussed proven to still hold. In this case, since the pressure ratio will vary from one stage to another, so will the adiabatic efficiency.

How the Impeller Produces Head

The foregoing discussion considered the polytropic head requirement necessary to elevate the pressure of the system. But how does the centrifugal impeller provide this head?

The centrifugal impeller imparts kinetic energy to the gas by increasing the gas velocity through the rotation of the impeller. Part of this kinetic energy results in a static pressure rise in the impeller; the remainder results in velocity head which later converts to an additional pressure rise in the diffuser located in the compressor stationary hardware. The kinetic energy is a function of the square of the velocity. Therefore, the head produced by an impeller is directly proportional to the square of the mechanical tip speed of the impeller:

$$H_p \propto u^2$$

where:

u = mechanical tip speed

Noting that the units of u^2 are ft^2/sec^2 (m^2/s^2), the units can be corrected to those of head—ft-lbf/lbm (N·m/kg)—by introducing the universal gravitational constant, g_c:

$$g_c = 32.2 \text{ ft-lbm/lbf-sec}^2 \ (1.0 \text{ kg·m/N·s}^2)$$

The proportion becomes:

$$H_p \propto \frac{u^2}{g_c}$$

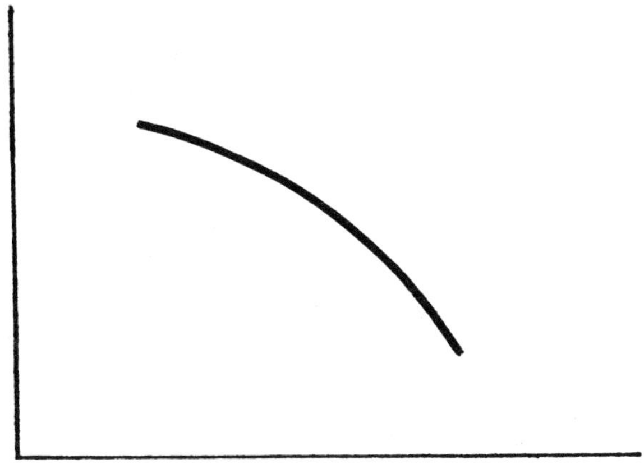

Figure 2-3. Typical variation of head coefficient with flow coefficient. (Note that the flow coefficient is a nondimensional flow term and a function of volume flow/rotational speed.)

Now introduce a nondimensional term, μ, defined as the head coefficient:

$$H_p = \frac{\mu u^2}{g_c} \tag{2-17}$$

$$H_p = \frac{\mu u^2}{32.2} \quad \text{in the English system}$$

$$H_p = \frac{\mu u^2}{1{,}000} \quad \text{in the metric system (in kN} \cdot \text{m/kg)}$$

The head coefficient, μ, is a variable for any particular impeller. It increases as the flow through the impeller decreases, and vice versa. The variation of μ with flow defines the characteristic curve of the impeller. A typical relationship is shown in Figure 2-3.

It is common practice in the industry to estimate that an impeller will nominally produce 10,000 ft-lbf/lbm (30 kN·m/kg) of polytropic head. This rule of thumb does not always apply, especially in very low-temperature or high-molecular-weight cases (this will be covered later). However, it may be interesting to note how the rule of thumb is derived. The head coefficient, μ, varies from about 0.4 to about 0.6. Assume an average value of μ =

(text continued on page 36)

Table 2-2
Centrifugal Compressor Equations—English Units

General

$$P_r = \frac{P}{P_c} \qquad T_r = \frac{T}{T_c} \qquad \frac{n}{n-1} = \frac{k}{k-1}\eta_p$$

Compressor Ratio Equations

Pressure Ratio	*Temperature Ratio*	*Volume Ratio*
$r_p = \frac{P_2}{P_1}$	$r_t = \frac{T_2}{T_1}$	$r_v = \frac{v_1}{v_2}$
$r_p = r_t^{n/(n-1)}$	$r_t = r_p^{(n-1)/n}$	$r_v = r_p^{1/n}$
$r_p = r_v^n$	$r_t = r_v^{n-1}$	$r_v = r_t^{1/(n-1)}$

Compressor Process Equations

	Polytropic	*Adiabatic*
Head:	$H_p = Z_a \frac{1{,}545}{MW} T_1 \frac{n}{n-1} [r_p^{(n-1)/n} - 1]$	$H_{ad} = Z_a \frac{1{,}545}{MW} T_1 \frac{k}{k-1} [r_p^{(k-1)/k} - 1]$
Discharge temperature:	$T_2 = r_t T_1 = r_p^{(n-1)/n}\, T_1$	$T_2 = \frac{T_1 [r_p^{(k-1)/k} - 1]}{\eta_{ad}} + T_1$
Power:	$PWR_g = \frac{mH_p}{33{,}000\ \eta_p}$	$PWR_g = \frac{mH_{ad}}{33{,}000\ \eta_{ad}}$

Units

p	psia	H	ft-lbf/lbm
T	°R	m	lbm/min
v	ft^3/lbm	PWR	horsepower
MW	lbm/lbm-mol	η	%

The following terms are dimensionless:

P_r	$k = \frac{c_p}{c_v}$	r_v
T_r	r_p	Z_a
n	r_t	

Table 2-3
Centrifugal Compressor Equations—Metric Units

General

$P_r = \dfrac{P}{P_c}$	$T_r = \dfrac{T}{T_c}$	$\dfrac{n}{n-1} = \dfrac{k}{k-1}\eta_p$

Compressor Ratio Equations

Pressure Ratio	***Temperature Ratio***	***Volume Ratio***
$r_p = \dfrac{P_2}{P_1}$	$r_t = \dfrac{T_2}{T_1}$	$r_v = \dfrac{v_1}{v_2}$
$r_p = r_t^{n/(n-1)}$	$r_t = r_p^{(n-1)/n}$	$r_v = r_p^{1/n}$
$r_p = r_v^n$	$r_t = r_v^{n-1}$	$r_v = r_t^{1/(n-1)}$

Compressor Process Equations

	Polytropic	***Adiabatic***
Head:	$H_p = \dfrac{Z_a}{1{,}000}\dfrac{8{,}314}{MW}T_1\dfrac{n}{n-1}[r_p^{(n-1)/n} - 1]$	$H_{ad} = \dfrac{Z_a}{1{,}000}\dfrac{8{,}314}{MW}T_2\dfrac{k}{k-1}[r_p^{(k-1)/k} - 1]$
Discharge temperature:	$T_2 = r_t T_1 = r_p^{(n-1)/n} T_1$	$T_2 = \dfrac{T_1[r_p^{(k-1)/k} - 1]}{\eta_{ad}} + T_1$
Power:	$PWR_g = \dfrac{mH_p}{3{,}600\ \eta_p}$	$PWR_g = \dfrac{mH_{ad}}{3{,}600\ \eta_{ad}}$

Units

p	kPa	H	kN·m/kg
T	K	m	kg/h
v	m^3/kg	PWR	kW
MW	kg/kg-mol	η	%

The following terms are dimensionless:

P_r	$k = \dfrac{c_p}{c_v}$	r_v
T_r	r_p	Z_a
n	r_t	

0.5. The nominal operating tip speed of an impeller is approximately 800 ft/sec (244 m/s). The polytropic head is then:

$$H_p = \frac{(0.5)(800)^2}{32.2} = 9{,}940 \approx 10{,}000 \text{ ft-lbf/lbm}$$

$$H_p = \frac{(0.5)(244)^2}{1{,}000} = 29.8 \approx 30 \text{ kN}\cdot\text{m/kg}$$

Summary

The reader has been familiarized with many equations in this chapter. The most important equations—those used in centrifugal compressor estimates—are presented in Tables 2-2 and 2-3 as a matter of convenience. The tables include all correction factors; the equations are therefore in usable form as they appear. The listed units are used extensively in the industry.

3
Gas Property Calculations

Calculating Gas Properties from a Gas Analysis

Once again, refer to the general form of the head equation to see what values are necessary to calculate the polytropic head:

$$H_p = ZRT_1 \frac{n}{n-1} [r_p^{(n-1)/n} - 1] \tag{2-12}$$

The values that must be known for this calculation are:

1. The inlet and discharge pressures for the pressure ratio.
2. The inlet temperature.
3. The polytropic efficiency.
4. The molecular weight, k-value, and compressibility of the gas or gas mixture under consideration. (Actually, the average k-value and average compressibility are needed for the head calculation for multistage compressors, and the average k-value and inlet compressibility for single-stage compressors. The explanation of this appears in Chapter 5, "Why Use Average k- and Z-Values?" For the purpose of this chapter, the reader should dispense with the average values and only concern himself with the method of calculating state-point k- and Z-values. Once developed, the procedure can be extended to a determination of average k- and Z-values.)

The inlet and discharge pressures and inlet temperature come from the process design. These values establish the process state points. The polytropic efficiency can be obtained either through estimating data presented in this book or from manufacturer's data. However, the molecular weight, k-value, and compressibility of the gas or gas mixture must be calculated.

The molecular weight will be a constant throughout compression except when interstage cooling results in liquid knockout. The k-value will be a function of the temperature and derives from the specific heat at constant pressure, c_p. The compressibility will be a function of both the temperature and the pressure, and derives from the reduced temperature and reduced pressure.

In order to calculate the molecular weight, k-value, and compressibility, it must be assumed that each gas contributes to the mixture behavior based on its mol fraction (volume fraction) of the mixture. For example, assume a two-component mixture. The mol fraction of the first component is x. The mol fraction of the other component is then $1 - x$. The molecular weight would be:

$$MW_{mix} = xMW_1 + (1 - x)MW_2$$

Likewise, each gas contributes to the mixture specific heat, critical pressure, and critical temperature based on its mol fraction:

$$c_{p_{mix}} = xc_{p_1} + (1 - x)c_{p_2}$$

$$P_{r_{mix}} = xP_{r_1} + (1 - x)P_{r_2}$$

$$T_{r_{mix}} = xT_{r_1} + (1 - x)T_{r_2}$$

Then, the mixture gas properties are obtained by summing the individual gas contributions.

The procedure used to obtain the gas properties is as follows:

1. List all the gases in the mixture along with their respective mol fractions.
2. List the molecular weights, critical pressures, and critical temperatures for each gas.
3. List the specific heats at constant pressure, c_p, for each gas, based on the temperature at which the mixture is measured (the specific heats must be on a mol basis, i.e. Btu/lbm-mol-°R or J/kmol·K).
4. Calculate and list the contribution of each gas to the mixture molecular weight by multiplying the mol fraction of each gas by its respective molecular weight.
5. Calculate and list the contribution of each gas to the mixture critical pressure, critical temperature, and specific heat by the same approach used in step 4 for the molecular weight.

6. Sum the individual gas contributions to obtain the mixture molecular weight.
7. Similarly, sum the individual contributions to obtain the mixture specific heat. The ratio of specific heats, k, results from one of the following equations:

$$k = \frac{c_p}{c_p - 1.986} \quad \text{in English system} \tag{3-1}$$

$$k = \frac{c_p}{c_p - 8.314} \quad \text{in metric system} \tag{3-2}$$

8. Sum the individual contributions to obtain the critical pressure and critical temperature of the mixture. These values are necessary to determine the compressibility. First, however, the reduced pressure and reduced temperature must be calculated:

$$P_r = \frac{P}{P_c} \tag{2-3}$$

$$T_r = \frac{T}{T_c} \tag{2-4}$$

where P and T are the pressure and temperature at which the gas is measured.

9. Obtain the compressibility from the compressibility chart now that P_r and T_r are known.

EXAMPLE 3-1

Consider a gas mixture whose molar composition is 80% propane, 15% n-butane, and 5% ethane. The mixture is at 30 psia and 60°F. What are the mixture molecular weight, k-value, and compressibility?

Data for the three gases in the example are given in Table 3-1. The data are listed in both the English and metric systems. More complete data are available in the literature and in Appendix C.

First calculate the contributions of propane to the molecular weight, critical pressure, critical temperature, and specific heat:

$$MW_{\text{due to propane}} = (0.8)(44.10) = 35.28$$

$$P_{C\text{due to propane}} = (0.8)\,(616) = 493$$

Table 3-1
Gas Properties (Partial Listing)

Gas	Hydrocarbon Reference Symbol	Chemical Formula	Molecular Weight	Critical Properties: English System P_c (psia)	English System T_c (°R)	Metric System P_c (kPa)	Metric System T_c (K)	C_p Btu/lbm-mol-°R at 60°F	C_p kJ/kmol·K at 15°C
Ethane	C_2	C_2H_6	30.07	708	550	4880	305	12.27	51.30
Propane	C_3	C_3H_8	44.10	616	666	4249	370	17.14	71.66
n-Butane	nC_4	C_4H_{10}	58.12	551	765	3797	425	22.96	96.01

Note: Values derived by permission from Gas Processors Suppliers Association, *Engineering Data Book*, 1972, Ninth Edition, 5th Revision, 1981; and GPSA *SI Engineering Data Book*, 1980.

Table 3-2
Mixture Property Calculation

Gas	Mol %	Molecular Weight	P_c (psia)	T_c (°R)	C_p Btu/lbm-mol-°R	Individual Gas Contributions: *MW*	P_c	T_c	C_p
Propane	80	44.10	616	666	17.14	35.28	493	533	13.71
n-Butane	15	58.12	551	765	22.96	8.72	83	115	3.44
Ethane	5	30.07	708	550	12.27	1.50	35	28	0.61
Total						45.50	611	676	17.76

$$T_{c\text{due to propane}} = (0.8)(666) = 533$$

$$c_{p\text{due to propane}} = (0.8)(17.14) = 13.71$$

Next calculate the contributions to the mixture properties due to the presence of *n*-butane and ethane. Table 3-2 is the result of the calculations.

Thus, the following mixture properties result:

$$MW_{\text{mix}} = 45.50$$

$$P_{c_{\text{mix}}} = 611$$

$$T_{c_{\text{mix}}} = 676$$

$$c_{p_{\text{mix}}} = 17.76$$

Calculate the ratio of specific heats for the mixture from Equation 3-1:

$$k_{\text{mix}} = \frac{c_{p_{\text{mix}}}}{c_{p_{\text{mix}}} - 1.986} = \frac{17.76}{17.76 - 1.986} = 1.126$$

Determine the compressibility by first calculating the reduced pressure and reduced temperature from Equations 2-3 and 2-4:

$$P_r = \frac{P}{P_c} = \frac{30}{611} = 0.049 \quad (2\text{-}3)$$

$$T_r = \frac{T}{T_c} = \frac{60 + 460}{676} = 0.769 \quad (2\text{-}4)$$

The compressibility comes from Figure C-1 now that P_r and T_r are known:

$$Z = 0.955$$

Thus the necessary mixture properties of the gas at 30 psia and 60°F are as follows:

$$MW = 45.50$$
$$k = 1.126$$
$$Z = 0.955$$

In order to determine the mixture properties, the mol fractions of each gas in the mixture must be known. Sometimes the gas analysis will be received in mols of constituent or in mass flow of constituent. These must be converted to mol fractions before the mixture properties can be calculated.

EXAMPLE 3-2

Given a mixture of 4,800 mols/hr of propane, 900 mols/hr of *n*-butane, and 400 mols/hr of ethane, what are the mol fractions of each of the constituents?

First determine the total mol flow of the mixture:

$$\text{total flow} = 4{,}800 + 900 + 400 = 6{,}100 \text{ mols/hr}$$

The mol fraction of each constituent equals the mol flow of the constituent divided by the total mol flow of the mixture. Therefore, for propane:

$$\text{mol\% propane} = \frac{4{,}800}{6{,}100}(100\%) = 78.7\%$$

Table 3-3 is the result of the calculations.

EXAMPLE 3-3

Given a mixture of 150,000 kg/h propane, 35,000 kg/h *n*-butane, and 15,000 kg/h ethane, what are the mol fractions of each of the constituents?

First, the mol flow must be determined from the mass flow. Then the mol fractions can be determined as in Example 3-2. The mol flow for each constituent is the mass flow of that constituent divided by the respective molecular weight. Therefore, for propane:

$$\text{mol flow propane} = \frac{150{,}000}{44.10} = 3{,}401 \text{ mols/h}$$

Table 3-4 is the result of the calculations.

Examples 3-2 and 3-3 demonstrate how to obtain the mol fractions. After calculating these, the mixture properties of the gas can be calculated.

Adding Water to Dry Gas

Assume a gas known to be saturated with water. Further, assume a fixed requirement on dry-gas flow through a compressor. In order to determine the mixture properties, the mol fraction of water vapor in the gas must be known. Furthermore, in order to determine the total flow through the compressor, the water content must be added to the required amount of dry gas. How is the water vapor in the gas computed?

Each gas contributes to the behavior of a mixture based on its mol fraction. Therefore, each gas contributes to the mixture pressure based on its

Table 3-3
Mol Fractions for Example 3-2

Constituent	Mol Flow (mol/hr)	Mol %
Propane	4,800	78.7
n-Butane	900	14.8
Ethane	400	6.5
Total	6,100	100.0

Table 3-4
Mol Fractions for Example 3-3

Gas	Mass Flow (kg/h)	Molecular Weight	Mol Flow (kmol/h)	Mol %
Propane	150,000	44.10	3,401	75.5
n-Butane	35,000	58.12	602	13.4
Ethane	15,000	30.07	499	11.1
Total	200,000	—	4,502	100.0

mol fraction. The mol fraction of each constituent is, then, the ratio of the partial pressure of that constituent to the total pressure of the mixture.

$$\text{mol fraction} = \frac{\text{partial pressure of constituent}}{\text{total pressure of mixture}}$$

Water can only exist in vapor form up to a definite partial pressure based on the temperature. This partial pressure is known as the saturation pressure at the temperature of concern. If the gas is saturated with water vapor, the mol fraction of water vapor will be equal to the ratio of the saturation pressure at the measured temperature to the total pressure of the mixture. Given that the mol fraction of water vapor is defined as y:

$$y = \frac{P_w}{P} \tag{3-3}$$

where:

y = mol fraction of water vapor
P_w = partial pressure of water vapor. It is exactly equal to the saturation pressure if the gas is saturated.

The mixture molecular weight can be determined from the following:

$$MW_{\text{mix}} = yMW_w + (1 - y)MW_{dg} \tag{3-4}$$

where:

w = water vapor
dg = dry gas

Further, the total mass flow can be determined from:

$$m_{\text{total}} = m_{dg}\left[1 + \frac{MW_w P_w}{MW_{dg} P_{dg}}\right] \tag{3-5}$$

Note that Equations 3-3, 3-4, and 3-5 hold true even if the water content is less than saturated. For these cases, the partial pressure of the water vapor must be determined by other means. It will be less than the saturation pressure. In an air-water vapor mixture, this can be handled through consideration of the relative humidity, which defines the actual water-vapor content as a percentage of the saturated water-vapor content at the same temperature.

EXAMPLE 3-4

Suppose one wishes to compress 1,000 lbm/min of dry carbon dioxide ($MW = 44.01$). The inlet pressure is 16 psia and the inlet temperature is 100°F. The gas is initially saturated with water vapor. What is the molecular weight of the mixture and the total mass flow for the required 1,000 lbm/min of dry CO_2 mass flow?

By reference to the Keenan and Keyes Steam Tables, the saturation pressure of water vapor at 100°F is 0.95 psia. Since the mixture is saturated, the partial pressure of water vapor is also 0.95 psia.

$$y = \frac{P_w}{P} = \frac{0.95}{16} = 0.059 = 5.9 \text{ mol\% water vapor}$$

$$\begin{aligned} MW_{\text{mix}} &= yMW_w + (1 - y)MW_{dg} \\ &= (0.059)(18.02) + (1 - 0.059)(44.01) \\ &= 42.48 \end{aligned}$$

$$\begin{aligned} m_{\text{total}} &= m_{dg}\left[1 + \frac{MW_w P_w}{MW_{dg} P_{dg}}\right] \\ &= 1{,}000\left[1 + \frac{(18.02)(0.95)}{(44.01)(16 - 0.95)}\right] \\ &= 1{,}026 \text{ lbm/min} \end{aligned}$$

The foregoing covered the first step in performing a compressor estimate—that of determining the mixture properties.

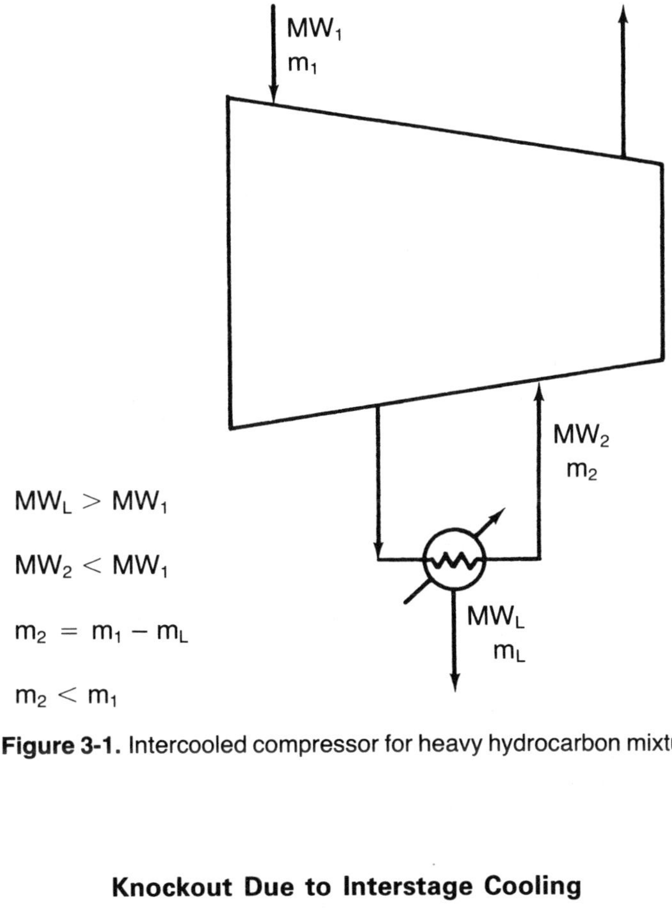

Figure 3-1. Intercooled compressor for heavy hydrocarbon mixtures.

Knockout Due to Interstage Cooling

Some compression processes require intermediate cooling to lower compression temperatures. The maximum operating temperature may be set by the compressor (e.g., metallurgy, expansion of components) or by the process (e.g., gas decomposition, fouling). Any time a gas is cooled, there is a potential for liquid knockout in the cooler. In the case of hydrocarbon mixtures, the liquid knockout is composed primarily of the heavier hydrocarbons, although, because of molecular interactions between the mixture constituents, some of the lighter hydrocarbons will leave with the liquid. The result of liquid knockout in hydrocarbon mixture compression will be that the mixture in the second section of the compressor will have a lower molecular weight than the mixture entering the first section of the compressor. The mass flow entering the second section will be lower also by the removal of the liquid. This is represented diagrammatically in Figure 3-1.

Calculating the second-section gas properties for hydrocarbon mixtures is very difficult and beyond the scope of this book. It is important, however. Remember that the polytropic head requirement is inversely proportional to the molecular weight. If the second-section polytropic head is calculated based on the first-section molecular weight ($MW_2 < MW_1$), the calculated value will inevitably be lower than actually required. Estimating the second section is further complicated by the mass-flow knockout. The knockout must be known in order to calculate the mass flow in the second section.

A dramatic example of this occurs when sizing a wet-gas compressor for a fluid catalytic cracker (FCC). The suction molecular weight is on the order of 35–45. Thus, the mixture contains a significant amount of heavy hydrocarbons. The intercool pressure is usually on the order of 65–75 psia (450–550 kPa). The amount of liquid knocked out in the intercooler is a function of pressure at the cool point. The higher the pressure, the more knockout, which implies lower molecular weights and lower mass flows. The mass flow entering the second section actually decreases more rapidly than the molecular weight. This can have a significant effect on power requirement. Therefore, when reviewing compressors for this type of process, precaution must be taken to ensure that manufacturers are compared on an equal basis.

As stated previously, the calculation of cooler knockout, and the resulting effects on downstream gas analyses, is beyond the scope of this book. Normally, the process department will determine by other means the gas analysis, mixture properties, and mass flow resulting from knockout. These are best presented as curves of molecular weight, k-value, compressibility, and mass flow versus intercool pressure. The reader should obtain this data for use in his compressor estimate. This data will also be necessary for communicating with the compressor vendor. If, however, compressor size and performance must be estimated without this information, assume that the inlet molecular weight, gas analysis, and mass flow apply throughout compression. The calculated head requirement will be low; however, the calculated power requirement will normally be conservative.

Water knockouts, on the other hand, are rather simple to handle. Gas properties and mass flows due to intercool water knockout can be determined by the same procedure used in Example 3-4.

4
Specifying Flow to a Compressor Manufacturer

There are basically three ways to specify compressor flow; all are used in the industry:

1. *Mass flow*—the amount of mass throughput per time element, normally lbm/min in the English system and kg/h in the metric system.
2. *Inlet, or actual, volume flow*—the amount of volume throughput per time element measured at the inlet to the compressor, normally cubic feet per minute (ICFM or ACFM) in the English system and cubic meters per hour (m^3/h) in the metric system.
3. *Standard volume flow*—the amount of volume throughput per time element, where the pressure and temperature used for measuring the volume are "standard" values. The units are normally standard cubic feet per minute (SCFM) or millions of standard cubic feet per day (MMSCFD) in the English system, and cubic meters per hour at 0°C ($m^3/h_{0°C}$) or cubic meters per hour at 15°C ($m^3/h_{15°C}$) in the metric system.

The advantages and disadvantages of each approach for specifying compressor flow are compared here.

Mass Flow

Mass flow is a specific value independent of gas properties or inlet conditions. For this reason, in the author's opinion, mass flow is the best way to specify compressor throughput when communicating with a compressor manufacturer.

There are some disadvantages to specifying mass flow, however. Manufacturers size their compressors based on volume flow. This makes sense, since the size of any container is a function of the volume of its contents. By specifying mass flow and not converting it to volume flow, an appreciation cannot be gained for the physical size of the compressor. As an extension of this, an impeller characteristic curve is a function of the inlet volume flow to the impeller. The impeller does not care how much mass it is compressing, but rather its volume. For instance, an impeller sized to pass 441,000 kg/h of propane (MW = 44.10) will pass only 20,200 kg/h of hydrogen (MW = 2.02) at the same pressure, temperature, speed, and polytropic head. (This assumes that the compressibilities are the same, also.)

It can be deduced from this that an impeller performance curve based on volume flow and polytropic head is insensitive to the inlet conditions (i.e. pressure, temperature, molecular weight, and compressibility). A performance curve based on mass flow is a direct function of the inlet conditions and is therefore only useful for the inlet conditions for which it was drawn. (For multi-stage compressors, varying inlet conditions can have some effect on the overall compressor performance curve, even when based on volume flow. This is due to the effect one stage has on another, known as volume ratio effects, which will be covered in more detail in Chapter 6. However, the foregoing is true for minor excursions of inlet conditions and will always be true for single-stage compressors.)

For some processes, there is another disadvantage to specifying mass flow: Confusion may result in gas streams saturated with water vapor. Water vapor is usually not part of the desired product. Sometimes the engineer is only interested in useful product flow and may specify the mass flow of the product less the water content, i.e. the dry-mass flow. The manufacturer must know water content to determine compressor size and performance. He may not be aware that the water content was not included. This disadvantage can be completely eliminated if all mass flows are specified as "wet" or "dry."

Even with the disadvantages discussed, there is less chance of confusion if mass flow is specified. This will become evident in the discussions that follow.

Inlet, or Actual, Volume Flow

Another method of specifying compressor flow is the inlet, or actual, volume flow to the compressor. The inlet volume flow is a function of the mass flow and the specific volume:

$$Q = mv \tag{4-1}$$

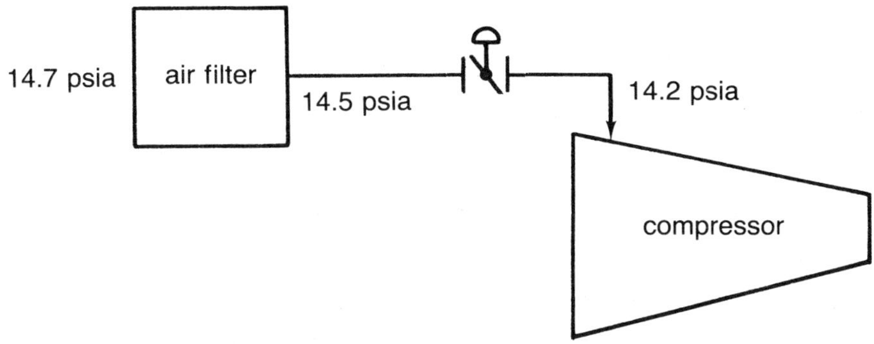

Figure 4-1. At which pressure was the compressor inlet volume specified?

where v is the specific volume and derives from one of the following:

$$v = \frac{ZRT}{144\,P} \quad \text{in the English system} \tag{4-2}$$

$$v = \frac{ZRT}{P \times 10^3} \quad \text{in the metric system, if } P \text{ is given in kPa} \tag{4-3}$$

This is a useful means of specifying compressor throughput, provided the pressure and temperature at which the gas is measured are clearly specified. In some cases, definition of the gas conditions may not be so obvious. In such cases, confusion may result from a specification of inlet volume flow. Consider Figure 4-1, which shows an air compressor with an inlet air filter and a butterfly valve. A volume flow of 10,000 ICFM (inlet cubic feet per minute) is specified. Which pressure was used for the specification? The potential confusion is obvious.

The advantage to communicating in inlet volume flow between disciplines or with the vendor in the estimating stages is an appreciation for the physical size of the compressor. The physical size affects compressor costs and may even dictate whether the application falls within the limits of a centrifugal compressor. However, mass flows must be converted to volume flows anyway in order to perform compressor estimates. So, while there is an advantage to communicating in terms of volume, there is really no advantage to using volume flow as the main throughput specification in a requisition or purchase order to a vendor. There is no problem in specifying both the mass flow and the volume flow on a data sheet; however, the vendor should be instructed to base his calculations and proposal on the mass flow.

A big disadvantage to volume flow specifications surfaces with a side-load refrigeration compressor. The volume coming through the side-load flange could be specified, but not the next section flow, since it is composed of the side-load flow and the previous section flow. The temperature at the point of mixing, which is actually the temperature into the next section, is a function of the previous section discharge temperature and, therefore, the previous section compression efficiency. Furthermore, the pressure at the mix point is a function of the pressure drop through the side-load flange, which only the manufacturer knows for sure.

While inlet volume flow is not the best means for specifying compressor throughput, it is the best means for plotting compressor performance curves. This is because the impeller, and therefore the compressor, is sensitive only to the actual volume throughput and is insensitive (except for volume ratio effects) to the gas state conditions. The reader should realize, therefore, that knowledge of the inlet volume flow is very important for gaining an appreciation for the compressor physical size and as a means of relating compressor performance.

Standard Volume Flow

The third method of specifying compressor flow is through the use of standard volume flow. This method is used extensively in the industry. In order to specify standard volume flow, the standard inlet conditions—pressure, temperature, molecular weight, and compressibility—must be known and held constant. Therefore, the standard specific volume is a constant:

$$v_{std} = \frac{Z_{std} R_{std} T_{std}}{P_{std}} = \text{constant}$$

Since the volume flow is given by:

$$Q_{std} = m v_{std}$$

the standard volume flow is directly proportional to the mass flow. Mass flow has therefore been specified (although the numerical value of the volume flow and the mass flow will obviously be different).

When specifying standard volume flow, an appreciation cannot be gained for the physical size of the compressor, since the numerical value of the standard volume flow can be orders of magnitude different from the numerical value of the inlet volume flow. Consider the following case. Say the standard pressure is 101.3 kPa. Assume that the actual inlet pressure under consideration is 1,000 kPa. The numerical value of the standard volume flow will be 10 times the numerical value of the inlet volume flow merely by ratio of the pressures. For this reason and for reasons discussed previously, the centrifugal impeller characteristic is not a function of standard volume flow.

Thus, the standard volume flow has none of the advantages of either the inlet volume flow or the mass flow. It does, however, normally resolve one of the disadvantages of the mass flow. It is normal practice in the industry for the standard volume flow to be based on dry gas. However, this practice is often abused, and one can't be sure when seeing standard volume flow whether it is wet or dry.

There is one serious disadvantage to using standard volume flow in a specification to a compressor manufacturer—that of confusion. There is no universally accepted set of standard conditions in either the English or metric system. Some of the more common "standard" conditions are as follows.

English system:

1. $P_{std} = 14.7$ psia
   $T_{std} = 60°F$
2. $P_{std} = 14.7$ psia
   $T_{std} = 70°F$
3. $P_{std} = 14.7$ psia
   $T_{std} = 32°F$

metric system:

1. $P_{std} = 101.3$ kPa
   $T_{std} = 0°C$
2. $P_{std} = 101.3$ kPa
   $T_{std} = 15°C$

The standard compressibility is normally 1.0, since most gases under these conditions would have $Z = 1.0$. The standard molecular weight is normally the molecular weight of the subject gas. The gas is normally assumed dry, although some of the "air" standards define the relative humidity at other than 0%.

As can be seen, there are several "standards" in common use today, varying from industry to industry and even from engineer to engineer. Unfortunately, even the rules of these standards are sometimes abused. The two most common abuses are the use of wet flow and the setting of standard compressibility at the value for the actual inlet conditions.

One must be very careful in using standard volume flow to define clearly the "standard" used. If standard volume flow is specified, it must be defined and this definition relayed to the compressor manufacturer.

Since standard volume flow is used extensively in the industry, one must know how to convert it for compressor estimates. Standard volume flow (SVF) can be converted to inlet volume flow (Q_1) by the following relationship in either the English or metric system:

$$Q_1 = SVF\left(\frac{Z_1}{Z_{std}}\right)\left(\frac{T_1}{T_{std}}\right)\left(\frac{P_{std}}{P_1}\right)\left(\frac{MW_{std}}{MW}\right) \tag{4-4}$$

where T is in absolute temperature—°R or K. Note that if SVF is based on dry gas, Q_1 will be also. Where appropriate, the water content must be added to the gas.

A better alternative to this approach would be to convert the standard volume flow directly to mass flow:

$$m = SVF\left(\frac{144\,P_{std}}{Z_{std}R_{std}T_{std}}\right) \quad \text{in the English system} \tag{4-5}$$

$$m = SVF\left(\frac{P_{std} \times 10^3}{Z_{std}R_{std}T_{std}}\right) \quad \text{in the metric system} \tag{4-6}$$

When using Equations 4-5 and 4-6, the time basis for m will be the same as the time basis for SVF; i.e. if SVF is on a per minute basis, m will be also. If SVF is expressed in millions of standard cubic feet per day (MMSCFD), Equation 4-5 is best used by first converting MMSCFD to SCFM (standard cubic feet per minute).

5

Centrifugal Compressor Estimates

Frame Data

Centrifugal compressor manufacturers do not have an infinite selection of hardware but rather standardize on a series of frame sizes that they choose from for all centrifugal compressor applications. In order to estimate compressor sizing, the frame data are needed. Often these data can be obtained from the manufacturers. If they are available, compressor sizing can be estimated fairly accurately; however, the estimate will only be valid for the particular manufacturer whose data were used in the analysis.

Lacking this information, Table 5-1 can be used, which lists six typical centrifugal compressor frame sizes covering nominal inlet volume flows from 1,000 to 100,000 ICFM (1,700 to 170,000 m^3/h). A few notes on Table 5-1 are needed:

1. The nominal flows listed in the table are inlet flows. As the gas proceeds through the compressor, its volume decreases such that the flow entering the second impeller is less than that entering the first, the third less than the second, and so on. Table 5-1 takes this factor into account: the reader can assume that for any inlet volume flow in the flow ranges listed, the manufacturer will have progressively smaller impellers (flow-wise) to satisfy compression requirements. (A current lower limit on impeller availability might be 200–300 ICFM (350–500 m^3/h). Thus, a compression process resulting in a discharge volume flow below 200 ICFM (350 m^3/h) will most likely not be a feasible application for a centrifugal compressor.)

Table 5-1
Typical Centrifugal Compressor Frame Data*

Frame	Nominal Inlet Volume Flow		Nominal Polytropic Head		Nominal Polytropic Efficiency (%)	Nominal Rotational Speed (RPM)	Nominal Impeller Diameter	
	English (ICFM)	Metric (m^3/h)	English (ft-lbf/lbm)	Metric (k·Nm/kg)			English (in)	Metric (mm)
A	1,000–7,000	1,700–12,000	10,000	30	76	11,000	16	406
B	6,000–18,000	10,000–31,000	10,000	30	76	7,700	23	584
C	13,000–31,000	22,000–53,000	10,000	30	77	5,900	30	762
D	23,000–44,000	39,000–75,000	10,000	30	77	4,900	36	914
E	33,000–65,000	56,000–110,000	10,000	30	78	4,000	44	1,120
F	48,000–100,000	82,000–170,000	10,000	30	78	3,300	54	1,370

*While this table is based on a survey of currently available equipment, the instance of any machinery duplicating this table would be purely coincidental.

2. Since cost increases with frame size, and giving consideration to item 1, the smallest frame size capable of handling the required inlet volume flow should always be selected unless a future plant expansion is anticipated. For example, if the process requirement was for 17,000 ICFM, Frame B should be selected unless a future plant expansion is desired; in which case Frame C should be selected.
3. Actual flow ranges for the various frame sizes available on the market vary from one manufacturer to another. The values tabulated in Table 5-1 should be regarded as typical only. Table 5-1 does result from a review of several manufacturers' data; however, any real machinery duplicating the table is purely coincidental.
4. The nominal inlet volume flows are what might be expected at the tabulated nominal speed. The fan law relationships, discussed later in this chapter, indicate that flow is directly proportional to speed: For applications requiring higher speeds, the flow range will increase proportionally; for applications requiring lower speeds, the flow range will decrease proportionally.
5. The actual polytropic efficiency will be a function of the actual impellers a manufacturer selects. Efficiencies will vary from impeller to impeller and from manufacturer to manufacturer. Those in Table 5-1 are typical for the frame sizes. For the purpose of this discussion, the polytropic efficiency is assumed a constant for each frame.
6. The nominal speed is a typically required rotational speed to produce 10,000 ft-lbf/lbm (30 k·Nm/kg) of polytropic head per stage.
7. The nominal impeller diameters in Table 5-1 are typical diameters for the nominal speeds and flow ranges listed. The reader will not use the diameters in the calculations; they are tabulated only to yield an appreciation for the physical size of the impeller.

Maximum Polytropic Head Per Stage

Table 5-1 indicates that the nominal polytropic head for a centrifugal impeller is 10,000 ft-1bf/1bm (30 k·Nm/kg). The actual nominal value for polytropic head will vary from manufacturer to manufacturer, and even from impeller to impeller, since it is a function of impeller geometry and the selected nominal speed. The actual variation will be from approximately 9,000 to 12,000 ft-lbf/lbm (27 to 36 k·Nm/kg). However, 10,000 ft-lbf/lbm (30 k·Nm/kg) is a good working number for estimating purposes.

Centrifugal compressor performance approximates a set of relationships commonly known as the fan laws. These laws state that the volume flow through an impeller is directly proportional to the speed of rotation:

$$Q \propto N$$

and

$$\frac{Q_n}{Q_o} = \frac{N_n}{N_o} \tag{5-1}$$

where:

N = rotational speed
n = new
o = old

Furthermore, the polytropic head produced by an impeller is proportional to the square of the rotational speed:

$$H_p \propto N^2$$

$$\frac{H_{p_n}}{H_{p_o}} = \left(\frac{N_n}{N_o}\right)^2 \tag{5-2}$$

Equation 5-2 indicates that if the speed is increased above nominal, the polytropic head produced is increased also. For example, take Frame A. The nominal speed of Frame A is 11,000 RPM. If the speed is increased to 12,000 RPM, the resultant polytropic head is:

$$\frac{H_{p_n}}{10{,}000} = \left(\frac{12{,}000}{11{,}000}\right)^2$$

$$H_{p_n} = 11{,}900 \text{ ft-lbf/lbm}$$

It would appear, then, that the faster an impeller spins, the more polytropic head produced. There is a practical limit, however. Impeller head production is primarily limited by two factors:

1. There is a limit to rotational speed, caused mainly by impeller stress levels and rotor critical speeds.
2. There is an aerodynamic limit to the mechanical tip speed associated with the proximity of gas velocity to unity Mach number.

Stress levels increase with increasing speed. At some point, the rotational speed results in stress levels approaching the yield stress of the impeller material. Obviously, operating at such speeds must be avoided. The yield stress of the impeller is a function of the material and the heat treatment used during fabrication. The actual stresses on the impeller are a function of its rotational speed and geometry. Thus, mechanical limitations will vary from manufacturer to manufacturer and from impeller to impeller, depending on material, heat treatment, and geometry. During estimating phases, the manufacturer's stress level data will not normally be available, so any useful criterion for mechanical speed limitations for estimating purposes cannot be derived from stress level considerations.

Rotor critical speeds are a function of impeller size and weight, shaft diameter, bearing size, and the bearing span (the length of the rotor between the bearings). All of these are variables, and manufacturers can alter rotor critical speeds, within limits, by varying one or more of them. Here again the manufacturer's critical speed data will not normally be available during the estimating phase, so useful criterion for mechanical speed limitations cannot be derived from considering rotor critical speeds.

One can, however, indirectly set limits on the rotational speed by reviewing its effect on impeller aerodynamics, thereby limiting the polytropic head per stage as a function of gas properties.

The performance of a centrifugal impeller is a function of its mechanical tip speed and the Mach number profile of the gas undergoing compression. The Mach number derives from Equation 5-3:

$$M = \frac{V}{a} \tag{5-3}$$

where:

M = Mach number
V = gas velocity
a = speed of sound of the gas

The speed of sound of the gas derives from Equation 5-4:

$$a = \sqrt{kg_c ZRT} \tag{5-4}$$

From this, it can be deduced that any two gases that have the same speed of sound and that are compressed in identical impellers at the same mechanical tip speed will yield the same performance.

Consider what happens as the mechanical tip speed of an impeller is increased with constant gas conditions, i.e. constant speed of sound. As speed increases, the lowest stable flow point (defined as the surge point) increases approximately as the fan law implies. The highest stable flow point (defined as the stonewall point) initially will also increase approximately with the fan law. However, at a certain speed, unity Mach number is approached somewhere in the impeller at a much lower flow than might be graphically anticipated. This limits the amount of flow that can pass through the impeller.

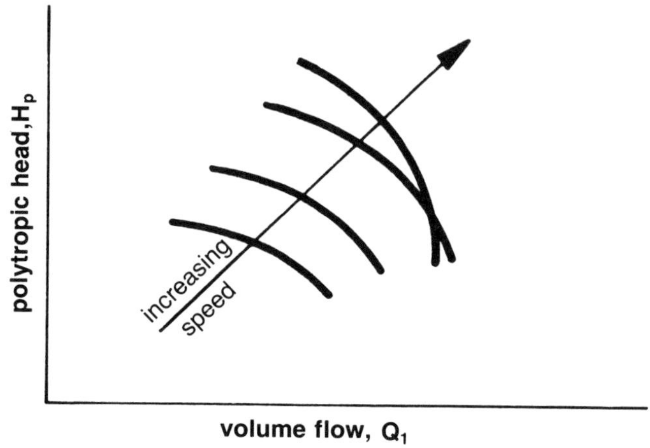

Figure 5-1. Typical impeller flow-head curve for variable speed.

The flow stability, then, becomes drastically reduced at higher speeds. A typical example is shown in Figure 5-1.

The foregoing discussion was based on a constant speed of sound. If the speed of sound of the gas decreases, it can be deduced from Equation 5-3 that the reduction in stable flow range will occur at a lower mechanical tip speed. Thus, the acceptable maximum mechanical tip speed and, therefore, the maximum rotational speed decrease with decreasing speed of sound of the gas. Also, since the polytropic head is related to the mechanical tip speed by Equation 2-17:

$$H_p = \frac{\mu u^2}{g_c} \tag{2-17}$$

the maximum polytropic head that can be expected from a centrifugal impeller decreases with decreasing speed of sound of the gas.

This analysis can be used to limit the rotational speed by limiting the maximum polytropic head per stage as a function of gas speed of sound.*

In order to set the limit, a datum point must first be established. Consider the datum point to be air at 80°F (27°C). Now define a speed-of-sound index, θ, which will be the square root of the ratio of the speed of sound in air at

*This procedure was originally developed by Elliott Company, Jeannette, PA. However, it can be used with any compressor and offers a good approach for limiting polytropic head per stage.

80°F (27°C) to the speed of sound in the gas under consideration. In the English system:

$$a_{\text{air},80°\text{F}} = \sqrt{kg_c ZRT} = \sqrt{(1.4)g_c(1.0)\left(\frac{1{,}545}{28.97}\right)(460 + 80)}$$

$$a_{\text{gas}} = \sqrt{kg_c ZRT} = \sqrt{k_{\text{gas}}\, g_c\, Z_{\text{gas}} \left(\frac{1{,}545}{MW_{\text{gas}}}\right) T_{\text{gas}}}$$

$$\theta = \frac{a_{\text{air},80°\text{F}}}{a_{\text{gas}}} = \sqrt{\left(\frac{1.4}{k_{\text{gas}}}\right)\left(\frac{g_c}{g_c}\right)\left(\frac{1.0}{Z_{\text{gas}}}\right)\left(\frac{1{,}545}{1{,}545}\right)\left(\frac{MW_{\text{gas}}}{28.97}\right)\left(\frac{540}{T_{\text{gas}}}\right)}$$

$$\theta = \sqrt{\frac{26.1\, MW_{\text{gas}}}{k_{\text{gas}}\, Z_{\text{gas}}\, T_{\text{gas}}}} \quad \text{in the English system} \qquad (5\text{-}5)$$

$$\theta = \sqrt{\frac{14.5\, MW_{\text{gas}}}{k_{\text{gas}}\, Z_{\text{gas}}\, T_{\text{gas}}}} \quad \text{in the metric system} \qquad (5\text{-}6)$$

Any two gases operating with the same speed of sound will yield the same performance. Therefore, the stability with a given gas operating at a given mechanical tip speed can be compared to the equivalent stability if the gas were air at 80°F (27°C):

$$u_e = u\,\theta \qquad (5\text{-}7)$$

where:

u_e = equivalent tip speed for air at 80°F (27°C)
u = actual mechanical tip speed
θ = square root of the ratio of the speeds of sound

For example, if the gas has $\theta = 1.3$ and operates at 600 fps (183 m/s) mechanical tip speed, it will yield the same stability as air at 80°F (27°C) operating at a mechanical tip speed of $1.3 \times 600 = 780$ fps (238 m/s). However, by Equation 2-17, the polytropic head produced would still only be a function of the actual mechanical tip speed of 600 fps (183 m/s).

By setting some conservative limits on maximum acceptable mechanical tip speed and equivalent tip speed and assuming a conservative value for the polytropic head coefficient, a curve of maximum polytropic head per stage versus θ can be developed. This has been done and is presented as Figures

5-2 and 5-3. On these curves, the subscript for the gas properties has been changed from "gas" to "1" to indicate inlet properties.

For most applications, the rotational speed required to produce the maximum polytropic head per stage as limited by Figures 5-2 and 5-3 will be well within the impeller yield stress and rotor critical speeds. One may wish to apply some additional conservatism when compressing gas mixtures requiring low-yield-strength material. Such mixtures would be corrosive gases requiring stainless-steel impellers and gas mixtures containing hydrogen sulfide. For these applications, one should limit the polytropic head per stage to the value from the curves or to 9,000 ft-lbf/lbm (27 k·Nm/kg), whichever is lower.

Maximum Number of Impellers Per Compressor Body

The maximum rotor length, and therefore the maximum number of impellers, is limited mainly by the rotor critical speeds. Most process centrifugals operate between the first and second rotor critical speeds. As the rotor is lengthened, both critical speeds decrease; however, the second critical speed decreases at a much faster rate than the first. As a result, the acceptable operating-speed range decreases with increasing rotor length. Rotor-length limits vary from manufacturer to manufacturer and possibly within a manufacturer's line. However, for the purpose of this discussion, the reader should consider that a straight-through compressor body can accommodate up to nine impellers.

For a single-cooled compressor, the intermediate diaphragm used to extract and reinject the gas often requires a space in the compressor body approximately equal to the axial length of an impeller. Therefore, for these applications, the maximum number of impellers should be reduced to eight in order not to exceed the nine-stage space limit.

For side-load applications, the side-load diaphragms and area alloted for mixing the two gas streams will normally require a stage space also. Thus, for side-load applications, the maximum number of impellers should be reduced by one for each side-load in the compressor.

Another type of arrangement is the double-flow compressor. With this arrangement the inlet flow is split and injected equally into both ends of the compressor. The gas is compressed in both halves of the casing and exits through a common discharge nozzle at the center of the compressor body. The impellers in both halves of the casing are identical. The effect is essentially to double the capacity of the compressor at the expense of one-half the potential head production. Since this arrangement requires two of each impeller, it cuts in half the maximum number of impellers per compressor

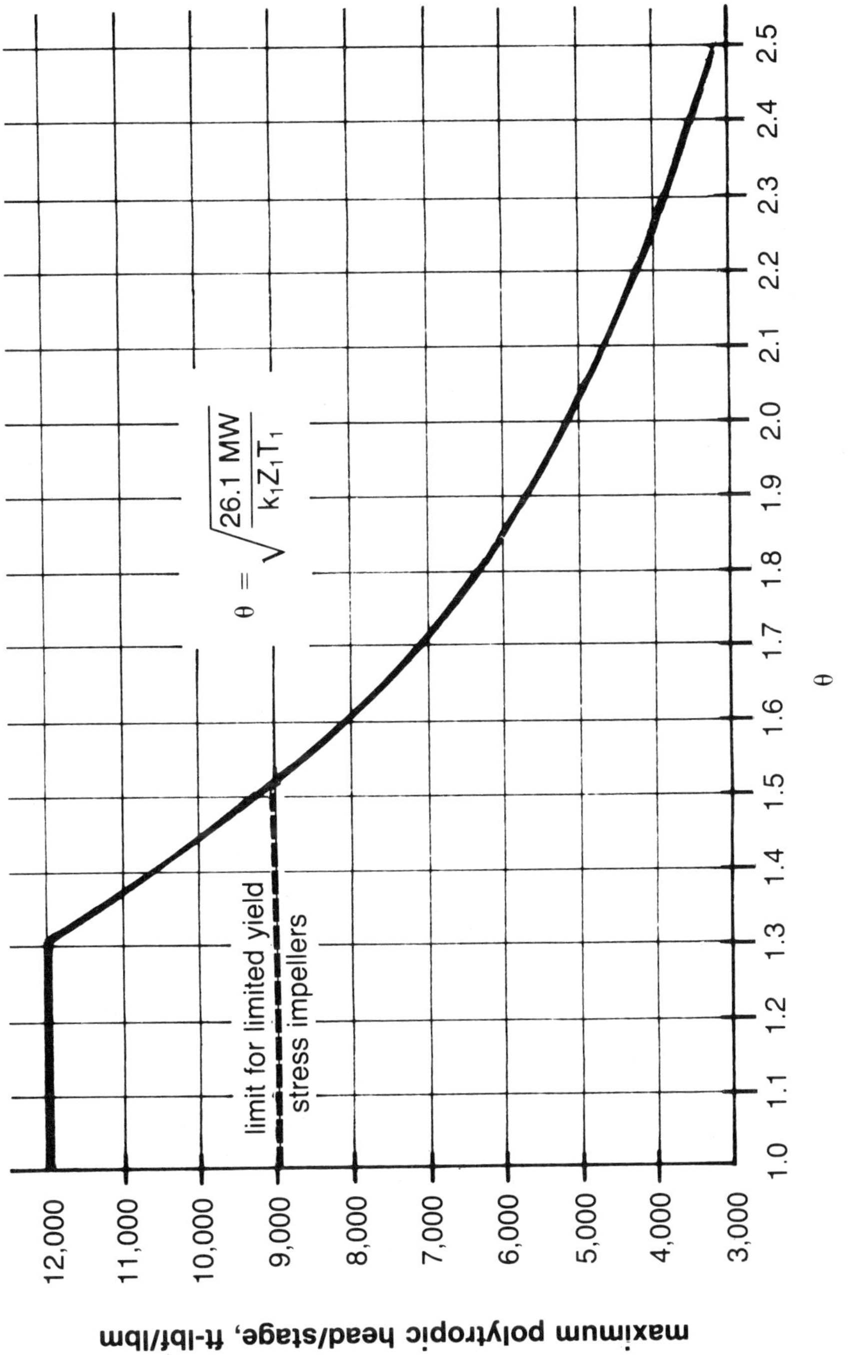

Figure 5-2. Maximum polytropic head per stage—English system.

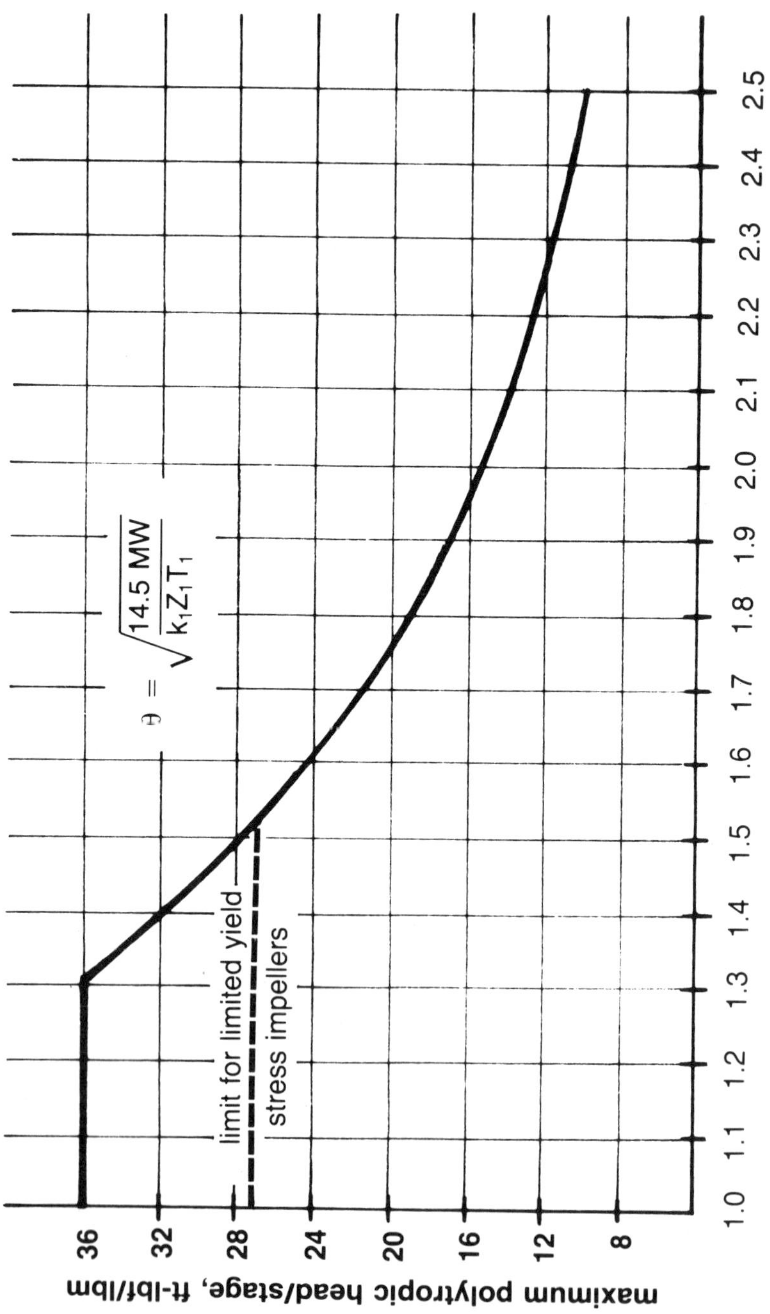

Figure 5-3. Maximum polytropic head per stage—metric system.

Table 5-2
Maximum Number of Impellers Per Compressor Body

Arrangement	Max No. Impellers/Body
Straight-through	9
Single intercooled	8
Double intercooled	7
1 side load or extraction	8
2 side loads or extractions	7
3 side loads or extractions	6
Double-flow	4

body (from the standpoint of head production). Therefore, the maximum number of impellers for a double-flow arrangement is four.

Table 5-2 summarizes the foregoing discussion.

Why Use Average *k*- and *Z*-Values?

Before investigating the compressor calculation procedures, it would be advantageous to review the merits of using average values of k and Z in calculating the polytropic head.

First, it should be realized that the inlet compressibility must be used to determine the actual volume flow entering the compressor, both to approximate the relative size of the compressor and as a means of communicating with the vender via data sheets. Furthermore, the maximum value of θ is of interest. θ will be at its maximum at the inlet to the compressor where the inlet compressibility occurs (although using the average compressibility will result in a conservative estimate of θ).

But what about the polytropic head? The compressibility continuously varies; in fact, it normally decreases while a gas undergoes compression. This would imply that use of the inlet compressibility would be conservative, since as the compressibility decreases, so does the head requirement for a specified pressure ratio (refer to Equations 2-15 and 2-16). However, if the variation in compressibility is significant, the polytropic head requirement calculated by using the inlet compressibility could be so conservative as to be practically useless when performing compressor calculations, and the results could not be used to check against vendor quotations. If compressor performance were to be calculated on a stage-by-stage basis as compressor manufacturers do, the inlet compressibility would be used for each stage. However, the inlet compressibilities would be different for each stage, depending on the inlet pressure and inlet temperature to that stage.

A very useful and accurate approximation can be substituted for the stage-by-stage technique by calculating the polytropic head for the overall section using the average compressibility. In effect, the head requirement is overestimated in the first half of the impellers and underestimated in the last half, thereby calculating a sectional polytropic head very near that calculated by the stage-by-stage technique. Note that since single-stage compressors only have one impeller, the inlet compressibility would be used for these applications.

Now look at the effects of the k-value (and therefore the n-value). The value of k also varies, and it, too, decreases during the compression process. Calculations should be performed with the average value of k, since the most accurate estimates on polytropic head and discharge temperature will result. (Actually, the average k-value should be used whether the calculation is stage-by-stage or sectional. Thus, calculations for the single-stage compressor should also be conducted with the average k-value.) An interesting note, however: The polytropic head equation is insensitive to k-value (and therefore to n-value) within the limits that k normally varies during compression. This is because any errors in the $n/(n - 1)$ multiplier in the polytropic head equation tend to balance corresponding errors in the $(n - 1)/n$ exponent. Therefore, a very useful approximation of polytropic head will result from using the inlet k-value. Discharge temperature, on the other hand, is very sensitive to k-value. Since the k-value normally decreases during compression, a discharge temperature calculated by using the inlet k-value will be conservative. But the calculated temperature may be several degrees higher than actual, possibly on the order of 25–50°F (15–30°C), for extreme cases. Should calculations be performed with the inlet k-value, this potential discrepancy should be kept in mind, especially when checking vendor quotations.

Calculating the average k-value can be time-consuming, especially for gas analyses containing several gases, since not only must the mol weighted c_p of the mixture be determined at the inlet temperature but also at the estimated discharge temperature. For extreme variations of k-value, one may even elect to iterate to find a good average k-value: estimating the discharge temperature from the inlet k-value, determining the value of k_2, using the average k-value to estimate a new discharge temperature, determining the new value of k_2, etc. This type of time-consuming fine-tuning is not necessary for compressor estimates; one would not entertain such measures in practice without the assistance of high-speed computers. The suggested approach is as follows:

1. If the k-value is felt to be highly variable, one pass should be made at estimating discharge temperature based on the inlet k-value; the average k-value should then be calculated using the estimated discharge temperature.

2. If the k-value is felt to be fairly constant, the inlet k-value can be used in the calculations.
3. If the k-value is felt to be highly variable, but sufficient time to calculate the average value is not available, the inlet k-value can be used (but be aware of the potential discrepancy in the calculated discharge temperature).

In the examples that follow, the inlet k-value will be used, since the primary purpose of the examples is to demonstrate centrifugal compressor estimating procedures.

A review of the polytropic head equation will show that, taking into account the preceding discussions, the inlet and discharge values are considered for each parameter except the temperature. So far it has been suggested that the polytropic head be calculated based on the inlet temperature. This is in fact the case. The inlet temperature is used to calculate polytropic head, not the average temperature. An appreciation for this will evolve from considering the discussion on adiabatic and polytropic processes in Chapter 2. When calculating compressor performance by the stage-by-stage technique, the manufacturer uses the inlet temperature to each stage to determine the discharge properties. Therefore, he considers the inlet temperature a variable in the technique. However, remember that with the polytropic process the sum of the individual stage heads equals the total sectional head. When the stage and sectional head requirements were calculated in Chapter 2, the inlet temperature was used in each case. Comparing the results proved that the sectional polytropic head should be based on sectional inlet temperature. If the adiabatic efficiency were independent of the pressure ratio, the same result would be obtained for adiabatic head. A note of caution: The polytropic head is based on the inlet temperature, provided there is no cooling. When the gas is cooled, the process must be broken into sections at each cool point, using the inlet temperature for each section in the calculations.

N-Method for Estimating Compressor Sizing

The first estimating procedure to investigate is the N-method. The procedure is so called due to the extensive use of the n-value in the calculations. Recall that the n-value is defined by Equation 2-13:

$$\frac{n}{n-1} = \frac{k}{k-1}\eta_p \tag{2-13}$$

This procedure should be used for any gas mixture for which a Mollier chart is not available.

The N-method calculation procedure follows:

1. Determine the gas mixture properties—molecular weight, critical pressure, critical temperature, and specific heat at constant pressure—as outlined in Chapter 3.
2. Determine the inlet compressibility by determining the inlet reduced pressure and reduced temperature as follows:

   a. $P_r = \dfrac{P_1}{P_c}$

   b. $T_r = \dfrac{T_1}{T_c}$

   c. Z_1 from compressibility charts in Appendix C.

3. Determine the inlet volume flow:

$$Q = mv \tag{4-1}$$

$$Q_1 = (m_1)\left(\frac{Z_1RT_1}{144P_1}\right) \quad \text{in English system} \tag{5-8}$$

$$Q_1 = (m_1)\left(\frac{Z_1RT_1}{P_1 \times 10^3}\right) \quad \text{in metric system} \tag{5-9}$$

where:

$$R = \frac{1{,}545}{MW} \quad \text{in English system}$$

$$R = \frac{8{,}314}{MW} \quad \text{in metric system}$$

4. Select the compressor frame using the calculated inlet volume flow from Table 5-1. Note that this table is typical; while it should reasonably resemble actual equipment, one should not expect to find identical compressor sizes or nomenclature in the industry.
5. Determine the pressure ratio from Equation 5-10:

$$r_p = \frac{P_2}{P_1} \tag{5-10}$$

6. Determine the approximate discharge temperature as follows:
   a. Determine the inlet k-value from Equation 3-1 or 3-2:

$$k = \frac{c_p}{c_p - 1.986} \quad \text{in English system} \tag{3-1}$$

$$k = \frac{c_p}{c_p - 8.314} \quad \text{in metric system} \tag{3-2}$$

   b. Determine the inlet n-value from Equation 2-13:

$$\frac{n}{n-1} = \frac{k}{k-1}\eta_p \tag{2-13}$$

   η_p comes from Table 5-1 for the chosen compressor frame.

   c. Determine the approximate discharge temperature from Equation 5-11:

$$T_2 = T_1(r_p)^{(n-1)/n} \tag{5-11}$$

7. Determine the average compressibility (Z_a):
   a. Z_1 from preceding step 2.
   b. Z_2 in like manner using P_2 and T_2.
   c. Z_a from:

$$Z_a = \frac{Z_1 + Z_2}{2}$$

8. Determine the average k-value (k_a):
   a. k_1 from preceding step 6.
   b. Determine specific heat at constant pressure, c_p, at T_2.
   c. Determine k_2 as k_1, but using c_p at T_2.
   d. k_a from:

$$k_a = \frac{k_1 + k_2}{2}$$

9. Determine the average value for $n/(n - 1)$ from the average k-value as in preceding step 6.

10. Determine the polytropic head from Equation 2-15 or 2-16:

    In English system:

$$H_p = Z_a \frac{1{,}545}{MW} T_1 \frac{n}{n-1} [r_p^{(n-1)/n} - 1] \tag{2-15}$$

    In metric system:

$$H_p = \frac{Z_a}{1{,}000} \frac{8{,}314}{MW} T_1 \frac{n}{n-1} [r_p^{(n-1)/n} - 1] \tag{2-16}$$

11. Determine the required number of compressor stages (impellers) as follows:
    a. Determine the value of θ from Equation 5-12 or 5-13:

$$\theta = \sqrt{\frac{26.1\ MW}{k_1\ Z_1\ T_1}} \quad \text{in English system} \tag{5-12}$$

$$\theta = \sqrt{\frac{14.5\ MW}{k_1\ Z_1\ T_1}} \quad \text{in metric system} \tag{5-13}$$

    Note that the inlet *k*- and *Z*-values are used for calculating θ. This is because the maximum value of θ occurs at the inlet. Therefore, the first impeller of a compressor body will set the maximum polytropic head per stage.
    b. Determine the maximum polytropic head per stage from either Figure 5-2 or 5-3.
    c. Determine the number of stages from Equation 5-14:

$$\text{no. stages} = \frac{H_p}{\text{max. } H_p/\text{stage}} \tag{5-14}$$

    Round off to the next higher whole number of stages.
12. Determine the required rotational speed from the fan law as presented in Equation 5-15:
    a. Nominal polytropic head (H_{nom}) from Table 5-1.
    b. Nominal rotational speed (N_{nom}) from Table 5-1.
    c. Required rotational speed from Equation 5-15:

$$N = N_{nom} \sqrt{\frac{H_p}{H_{p_{nom}} \times \text{no. stages}}} \tag{5-15}$$

Table 5-3
Approximate Mechanical Losses as a Percentage of Gas Power Requirement*

Gas Power Requirement		Mechanical
English (hp)	Metric (kW)	Losses, L_m (%)
0–3,000	0–2,500	3
3,000–6,000	2,500–5,000	2.5
6,000–10,000	5,000–7,500	2
10,000+	7,500+	1.5

*There is no way to estimate mechanical losses from gas power requirements. This table will, however, ensure that mechanical losses are considered and yield useful values for estimating purposes.

13. Determine the required shaft power as follows:
    a. Determine the gas power requirement from either Equation 5-16 or 5-17:

$$PWR_g = \frac{mH_p}{33{,}000\ \eta_p} \quad \text{in English system} \tag{5-16}$$

$$PWR_g = \frac{mH_p}{3{,}600\ \eta_p} \quad \text{in metric system} \tag{5-17}$$

    b. Determine the mechanical losses from Table 5-3, based on the gas power calculated in step 13a.
    c. Determine the required shaft power from Equation 5-18:

$$PWR_s = PWR_g + L_m \tag{5-18}$$

14. Determine the discharge temperature as in step 6 except use the average *n*-value:

$$T_2 = T_1(r_p)^{(n-1)/n} \tag{5-11}$$

By this procedure, the compressor can be completely defined: The inlet volume flow, polytropic head requirement, shaft horsepower, and discharge temperature can be calculated; the compressor frame size, the required number of compressor stages, and the rotational speed of operation can be determined.

EXAMPLE 5-1

Consider a gas mixture whose molar composition is 80% propane, 15% *n*-butane, and 5% ethane, initially at 30 psia and 60°F. The object is to compress 5,000 lbm/min of the gas to 100 psia. Determine the compressor frame size, number of stages, rotational speed, power requirement, and discharge temperature.

1. Determine the gas properties (this same gas mixture was considered in Example 3-1):

$$MW = 45.5$$
$$P_c = 611 \text{ psia}$$
$$T_c = 676°R$$
$$c_p = 17.76$$
$$k_1 = 1.126$$

2. Determine the inlet compressibility. From Example 3-1:

$$Z_1 = 0.955$$

3. Determine the inlet volume flow:

$$Q_1 = m_1\left(\frac{Z_1RT_1}{144P_1}\right) = (5{,}000)\left[\frac{(0.955)(1{,}545)(60 + 460)}{(45.5)(144)(30)}\right]$$

$$Q_1 = 19{,}517 \text{ ICFM}$$

4. Select a compressor frame from Table 5-1. The smallest frame capable of the inlet volume flow is Frame C, with the following nominal data:

$$H_{p\text{nom}} = 10{,}000 \text{ ft-lbf/lbm}$$
$$\eta_p = 77\%$$
$$N_{\text{nom}} = 5{,}900 \text{ RPM}$$

5. Determine the pressure ratio:

$$r_p = \frac{P_2}{P_1} = \frac{100}{30} = 3.33$$

6. Determine the approximate discharge temperature:

$$k_1 = 1.126$$

$$\frac{n}{n-1} = \frac{k}{k-1}\eta_p = \frac{1.126}{0.126}(0.77) = 6.88$$

$$T_2 = T_1(r_p)^{(n-1)/n} = (60 + 460)(3.33)^{1/6.88} = 619°\text{R} = 159°\text{F}$$

7. Determine the average compressibility:

$$Z_1 = 0.955$$

$$(P_r)_2 = \frac{P_2}{P_c} = \frac{100}{611} = 0.164$$

$$(T_r)_2 = \frac{T_2}{T_c} = \frac{619}{676} = 0.916$$

$Z_2 = 0.925$ from compressibility charts in Appendix C

$$Z_a = \frac{Z_1 + Z_2}{2} = \frac{0.955 + 0.925}{2} = 0.94$$

8. Determine the average k-value. As previously discussed, the inlet k-value will be used for simplicity:

$$k_a = 1.126$$

9. Determine the average $n/(n - 1)$ value from the average k-value. As with step 8:

$$\frac{n}{n-1} = 6.88$$

10. Determine the polytropic head:

$$H_p = Z_a R\, T_1 \frac{n}{n-1}\left[r_p^{(n-1)/n} - 1\right]$$

$$= (0.94)\left(\frac{1{,}545}{45.5}\right)(520)(6.88)[(3.33)^{1/6.88} - 1]$$

$$= 21{,}800 \text{ ft-lbf/lbm}$$

11. Determine the required number of compressor stages:

$$\theta = \sqrt{\frac{26.1\, MW}{k_1\, Z_1\, T_1}} = \sqrt{\frac{(26.1)(45.5)}{(1.126)(0.955)(520)}} = 1.46$$

max. H_p/stage = 9,700 ft-lbf/lbm from Figure 5-2, using $\theta = 1.46$

$$\text{no stages} = \frac{H_p}{\text{max. } H_p/\text{stage}} = \frac{21{,}800}{9{,}700} = 2.25 \rightarrow 3$$

12. Determine the required rotational speed:

$$N = N_{\text{nom}} \sqrt{\frac{H_p}{H_{p_{\text{nom}}} \times \text{no. stages}}}$$

$$= 5{,}900 \sqrt{\frac{21{,}800}{(10{,}000)(3)}} = 5{,}030 \text{ RPM}$$

13. Determine the required shaft power:

$$PWR_g = \frac{m\,H_p}{33{,}000\ \eta_p} = \frac{(5{,}000)(21{,}800)}{(33{,}000)(0.77)} = 4{,}290 \text{ hp}$$

Mechanical losses (L_m) are 2.5% from Table 5-3.

$$L_m = (.025)(4{,}290) = 107 \text{ hp}$$

$$PWR_s = PWR_g + L_m = 4{,}290 + 107 \approx 4{,}400 \text{ hp}$$

14. Determine the actual discharge temperature:

$$T_2 = T_1(r_p)^{(n-1)/n} = 520(3.33)^{1/6.88} = 619°\text{R} = 159°\text{F}$$

The discharge temperature from step 14 is the same as that calculated in step 6, only because of the decision to use the inlet k-value instead of the average k-value. Had the average k-value been used, the actual discharge temperature would have come out lower.

Thus, the selection is a Frame C centrifugal compressor with three impellers operating at 5,030 RPM, requiring 4,400 hp, with a discharge temperature of 159°F.

EXAMPLE 5-2

Reconsider the previous example, except that this time use Figures C-5 and C-10 in Appendix C as a quick approach to estimating k- and Z-values. This will also serve as a comparison of results.

From Example 5-1:

MW = 45.5
P_1 = 30 psia
P_2 = 100 psia
T_1 = 60°F
T_2 = 159°F (estimated from step 6, Example 5-1)

From Figure C-5, one can estimate the inlet and discharge values for compressibility and thereby calculate the average value:

$$MW = 45.5;\ P_1 = 30 \text{ psia};\ T_1 = 60°\text{F} \rightarrow Z_1 = 0.953$$

$$MW = 45.5;\ P_2 = 100 \text{ psia};\ T_1 = 159°\text{F} \rightarrow Z_2 = 0.915$$

$$Z_a = \frac{Z_1 + Z_2}{2} = \frac{0.953 + 0.915}{2} = 0.934$$

From Figure C-10, one can estimate the average k-value from the average temperature. (An average k-value thus obtained will closely approximate one obtained from averaging the inlet and discharge k-values. Furthermore, the calculated polytropic head is insensitive to small changes in k-value.)

$$T_a = \frac{T_1 + T_2}{2} = \frac{60 + 159}{2} = 110°\text{F}$$

$$MW = 45.5;\ \text{T} = 110°\text{F} \rightarrow k_a = 1.115$$

Before proceeding with the example, it might be a good idea to check out the accuracy of Figure C-10. Note that the inlet k-value was used in Example 5-1. If Figure C-10 is used to determine the inlet k-value:

$$MW = 45.5;\ T = 60°\text{F} \rightarrow k_1 = 1.122$$

which is very close to the value of $k = 1.126$ calculated by conventional means.

It would also be interesting to determine the average k-value based on k_1 and k_2 from Figure C-10 and compare it to that determined from Figure C-10 using the average temperature:

$$MW = 45.5;\ T = 60°\text{F} \rightarrow k_1 = 1.122$$

$$MW = 45.5;\ T = 159°\text{F} \rightarrow k_2 = 1.106$$

$$k_a = \frac{k_1 + k_2}{2} = \frac{1.122 + 1.106}{2} = 1.114$$

which is very close to $k_a = 1.115$ previously determined from the average temperature.

The previous calculations yielded:

$$MW = 45.5$$
$$k_a = 1.115$$
$$Z_a = 0.934$$

Calculate the value of $n/(n - 1)$:

$$\frac{n}{n-1} = \frac{k}{k-1}\eta_p = \frac{1.115}{0.115}(0.77) = 7.47$$

Calculate the polytropic head:

$$H_p = Z_a R\, T_1 \frac{n}{n-1}\left[r_p^{(n-1)/n} - 1\right]$$

$$= (0.934)\left(\frac{1{,}545}{45.5}\right)(520)(7.47)[(3.33)^{1/7.47} - 1] = 21{,}500 \text{ ft-lbf/lbm}$$

Compare this to the 21,800 ft-lbf/lbm calculated in Example 5-1, step 10. The difference from using the charts is only 1.4%. About half this difference is due to the inaccuracy of Figure C-5; the other half is due to the fact that here the average k-value was used, while in Example 5-1 the inlet k-value was used.

From here on, the method will be the same as that used in Example 5-1. Look at the discharge temperature, however, to compare the value calculated using the average k-value to that calculated using the inlet k-value. Using k_a from Figure C-10:

$$T_2 = T_1(r_p)^{(n-1)/n} = 520\ (3.33)^{1/7.47} = 611°\text{R} = 151°\text{F}$$

Recall that the discharge temperature calculated using the inlet k-value in Example 5-1 was 159°F. The difference in k values of $k_a = 1.115$ and $k_a = 1.126$ (inlet k-value used as average in Example 5-1) resulted in a difference in calculated discharge temperatures of approximately 8°F.

Interstage Cooling

When compression temperatures get too high, an intercooled compressor can often be employed. This type of compressor is actually two or three compressors in one body. Gas is removed from the compressor part way through the compression process, directed through an external heat exchanger, and reinjected into the compressor for the final compression.

While centrifugal compressors are capable of operating with discharge temperatures as high as 450°F (230°C), *a discharge temperature of 375°F (190°C) will be considered maximum for normal operation* unless process considerations require a lower maximum temperature.

An intercooled compressor can be sized with the same procedure outlined previously; however, an optimum point must be determined for the intercooling.

Optimum design for intercooled compressors occurs when all sectional discharge temperatures are approximately the same. A good rule of thumb would be to select the compressor such that the discharge temperatures are within 10°F (5°C) of each other.

Most often, intercooled compressors have the same inlet temperatures at each section. With many processes employing intercooled compressors, the process gas is cooled against water in a heat exchanger upstream of the compressor inlet and also against water in the intercool heat exchanger. Since the water source for both heat exchangers is the same, and since it is not a bad assumption that both heat exchangers will have the same approach temperature, it can be deduced that the process gas will have the same temperature at both compressor inlets. Furthermore, consider the intercooled air compressor. Ambient air enters the inlet of the compressor. To some degree the temperature of the cooling-water source for the intercool heat exchanger will follow the ambient air temperature. The cooling water, therefore, will normally be capable of cooling the gas in the intercooler reasonably close to ambient air temperature, at least when operating at or near the compressor rated conditions. The result is a fairly equal inlet temperature to both sections.

Now, since reasonably equal discharge temperatures are sought from each section, and since the sectional inlet temperatures are reasonably equal, it follows that optimum compressor design will evolve when the temperature ratios are the same for each section. Furthermore, with equal temperature distributions in the sections the k-values will be equal and, assuming constant polytropic efficiency, so will the n-values. Reference to Equation 2-8 will show that with equal temperature ratios and equal n-values the result is equal pressure ratios. This principle will establish the starting point for estimating the intersection discharge pressure for intercooled compressors.

Consider Figure 5-4, which schematically shows a single intercooled compressor. The overall pressure ratio for the compressor is given by:

$$r_p = \frac{P_2}{P_1}$$

If, for the time being, the reader ignores pressure drops due to the interstage cooler, piping, and fittings:

$$P_2' = P_1' = P'$$

and:

$$\frac{P_2}{P_1} = \frac{P'}{P_1} \times \frac{P_2}{P'}$$

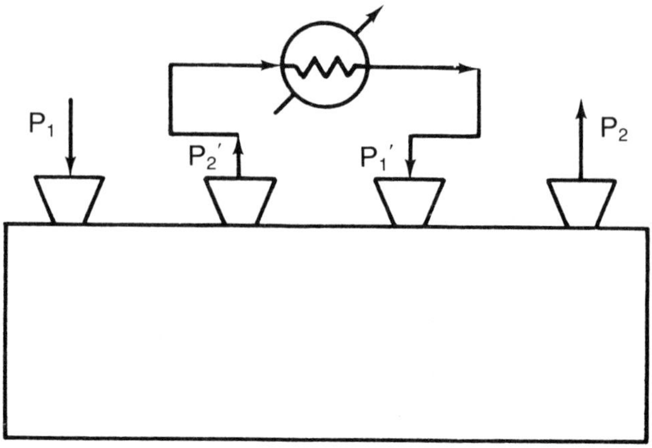

Figure 5-4. Schematic representation of a single intercooled compressor.

By definition:

$$(r_p)_{\text{SECT 1}} = \frac{P'}{P_1}$$

$$(r_p)_{\text{SECT 2}} = \frac{P_2}{P'}$$

Attempting to make the pressure ratios equal:

$$(r_p)_{\text{SECT 1}} = (r_p)_{\text{SECT 2}} = r_{p\text{SECT}}$$

Then:

$$r_p = r_{p\text{SECT}} \times r_{p\text{SECT}} = (r_{p\text{SECT}})^2$$

Therefore:

$$r_{p\text{SECT}} = \sqrt{r_p} \qquad (5\text{-}19)$$

A more general equation for determining the sectional pressure ratio for a compression duty requiring b sections is:

$$r_{p\text{SECT}} = \sqrt[b]{r_p} \qquad (5\text{-}20)$$

where:

r_p = overall pressure ratio

$r_{p_{SECT}}$ = sectional pressure ratio

b = number of compression sections

Thus, to use the N-method to size an intercooled compressor, the general technique need only be modified by adding the following steps:

1. Estimate the intercool pressure based on Equation 5-20.
2. Estimate the discharge temperatures of each section, incorporating known pressure drops in the heat exchanger and associated piping. (Note that when the pressure drops are incorporated, the pressure ratios will no longer be equal.)
3. Check the discharge temperatures to be sure they are less than 375°F (190°C) or less than the temperature limitations set by the process. If not, one or more sections will have to be added to satisfy maximum temperature requirements.
4. By trial and error, modify the intercool pressure to obtain discharge temperatures equal to within 10°F (5°C).

EXAMPLE 5-3

Consider a dry-air process with an inlet pressure of 100 kPa and inlet temperature of 34°C. The object is to compress 68,000 kg/h of the air to 700 kPa. Assume that the heat-exchanger pressure drop is 2% of the entering pressure, and that the gas reentry temperature is 38°C. Determine the compressor frame size, number of stages, rotational speed, power requirement, and discharge temperature.

Dry air has been selected purposely for simplicity. Dry air has a molecular weight of 28.96. If the air contained moisture, the water content would lower the molecular weight. The molecular weight could be determined with the procedures outlined in Chapter 3 or by referring to any of the thermodynamics books in the literature. Since the purpose here is to demonstrate the procedures for compressor selection, the exercise will not be burdened with a study of gas properties calculations. One note of caution, however: In practice, air will contain moisture. If the moisture content is high enough—and it usually is—water knockout will be experienced in the intercooler. All this means is that the mass flow and molecular weight into the second section must be determined, since they will be different than those entering the first section. The difference will be a direct function of the intercool pressure: the higher the intercool pressure, the more the water knockout and, as a result, the lower the mass flow and the higher the molecular weight into the second section.

The compressibility and ratio of specific heats do not vary much for air. Therefore, the following can always be used for estimates:

$$k = 1.4$$
$$Z = 1.0$$

Determine the inlet volume flow:

$$Q_1 = m_1\left(\frac{Z_1RT_1}{P_1 \times 10^3}\right) = (68{,}000)\left[\frac{(1.0)(8{,}314)(34 + 273)}{(28.96)(100 \times 10^3)}\right] = 59{,}932 \text{ m}^3/\text{h}$$

Select the compressor frame from Table 5-1. The smallest frame capable of the inlet volume flow is Frame D, with the following nominal data:

$$H_{p\text{nom}} = 30 \text{ k·Nm/kg}$$
$$\eta_p = 77\%$$
$$N_{\text{nom}} = 4{,}900 \text{ RPM}$$
$$d_{\text{nom}} = 914 \text{ mm}$$

Determine the pressure ratio:

$$r_p = \frac{P_2}{P_1} = \frac{700}{100} = 7.0$$

Determine approximate discharge temperature and, correspondingly, the number of compression sections required:

$$k = 1.4$$

$$\frac{n}{n-1} = \frac{k}{k-1}\,\eta_p = \frac{1.4}{0.4}(0.77) = 2.69$$

Assuming one section:

$$T_2 = T_1(r_p)^{(n-1)/n} = 307\ (7)^{1/2.69} = 633 \text{ K} = 360°\text{C}$$

The discharge temperature exceeds the maximum of 190°C; thus intercooling will be required. Therefore, assume two sections:

$$r_{p\text{SECT}} = \sqrt{r_p} = \sqrt{7} = 2.65$$

$$T_2 = T_1(r_p)^{(n-1)/n} = 307\ (2.65)^{1/2.69} = 441 \text{ K} = 168°\text{C}$$

which is a satisfactory discharge temperature.

Determine the average compressibility:

$$Z = 1.0$$

Determine the average k-value:

$$k = 1.4$$

Determine the average value for $n/(n - 1)$:

$$\frac{n}{n-1} = 2.69$$

Calculate the polytropic head. A polytropic head must be calculated for each section. To do this, the pressure profile and the resulting actual pressure ratios for the two sections must first be determined. In practice, the first section of the compressor will normally be more efficient than the second section, since the individual impellers used in the first section will usually have a higher efficiency. Therefore, it is advantageous to favor the first section when determining the pressure profile. For this reason, the intercool pressure drop has been added to the first section instead of the second.

$$P_1 = 100 \text{ kPa}$$

$$P_2' = r_{pSECT} \times P_1 \times 1.02 = (2.65)(100)(1.02) = 270 \text{ kPa}$$

$$P_1' = \frac{P_2'}{1.02} = \frac{270}{1.02} = 265 \text{ kPa}$$

$$P_2 = 700 \text{ kPa}$$

$$(r_p)_1 = \frac{P_2'}{P_1} = \frac{270}{100} = 2.7$$

$$(r_p)_2 = \frac{P_2}{P_1'} = \frac{700}{265} = 2.64$$

$$H_p = \frac{Z_a}{1{,}000}\,\frac{8{,}314}{MW}\,T_1\,\frac{n}{n-1}\,[r_p{}^{(n-1)/n} - 1]$$

$$(H_p)_1 = \left(\frac{1}{1{,}000}\right)\left(\frac{8{,}314}{28.96}\right)(34 + 273)(2.69)[(2.7)^{1/2.69} - 1] = 106 \text{ k·Nm/kg}$$

$$(H_p)_2 = \left(\frac{1}{1{,}000}\right)\left(\frac{8{,}314}{28.96}\right)(38 + 273)(2.69)[(2.64)^{1/2.69} - 1] = 104 \text{ k·Nm/kg}$$

Determine the required number of compressor stages:

$$\theta = \sqrt{\frac{14.5\,MW}{k_1\,Z_1\,T_1}} = \sqrt{\frac{(14.5)(28.96)}{(1.4)(1.0)(307)}} = 0.99$$

max. H_p/stage = 36 k·Nm/kg from Figure 5-3, with $\theta = 0.99$

$$(\text{no. stages})_1 = \frac{(H_p)_1}{\text{max. } H_p/\text{stage}} = \frac{106}{36} = 2.94 \rightarrow 3$$

$$(\text{no. stages})_2 = \frac{(H_p)_2}{\text{max. } H_p/\text{stage}} = \frac{104}{36} = 2.89 \rightarrow 3$$

Note that the total number of impellers will be six, which is within the guidelines established for a single intercooled compressor in Table 5-2.

Determine the required rotational speed. The speed for each section must be calculated; the higher value will be used.

$$N_1 = N_{\text{nom}} \sqrt{\frac{(H_p)_1}{H_{p_{\text{nom}}} \times \text{no. stages}}} = 4{,}900 \sqrt{\frac{106}{30 \times 3}} = 5{,}320 \text{ RPM}$$

$$N_2 = N_{\text{nom}} \sqrt{\frac{(H_p)_2}{H_{p_{\text{nom}}} \times \text{no. stages}}} = 4{,}900 \sqrt{\frac{104}{30 \times 3}} = 5{,}270 \text{ RPM}$$

Therefore, N = 5,320 RPM.

Determine the required shaft power:

first section:

$$PWR_g = \frac{m_1(H_p)_1}{3{,}600\,\eta_p} = \frac{(68{,}000)(106)}{(3{,}600)(0.77)} = 2{,}600 \text{ kW}$$

second section:

$$PWR_g = \frac{m_2(H_p)_2}{3{,}600\,\eta_p} = \frac{(68{,}000)(104)}{(3{,}600)(0.77)} = 2{,}550 \text{ kW}$$

$$PWR_g = 2{,}600 + 2{,}550 = 5{,}150 \text{ kW}$$

$$L_m = (.02)(5{,}150) = 103 \text{ kW}$$

$$PWR_s = PWR_g + L_m = 5{,}150 + 103 \approx 5{,}255 \text{ kW}$$

Determine the actual discharge temperature for each section:

first section: $T_2' = T_1(r_p)_1^{(n-1)/n} = 307(2.7)^{1/2.69} = 444\text{ K} = 171°\text{C}$

second section: $T_2 = T_1'(r_p)_2^{(n-1)/n} = 311(2.64)^{1/2.69} = 446\text{ K} = 173°\text{C}$

Both sectional discharge temperatures are relatively the same, therefore the selection is adequate.

Thus, the selection is a Frame D centrifugal compressor with a total of six impellers, three in each section, operating at 5,320 RPM, requiring 5,255 kW, with sectional discharge temperatures of 171°C and 173°C, respectively.

So far, only applications where the inlet temperatures are fairly equal have been considered. This will not always be the case; however, the reader should always try to optimize at equal discharge temperatures. A rather dramatic example of non-equal inlet temperatures for an intercooled compressor is the chlorine application. For this process, the inlet temperature to the compressor is normally in the range of 0°F to −30°F (−20°C to −35°C). The intercool heat exchanger usually uses water as the cooling medium. Therefore, the second section inlet temperature will be in the range of 95–105°F (35–40°C). Due to fouling, process considerations will limit the discharge temperature to the 225–275°F (105–135°C) range. The inlet temperatures are quite different. The result will be to raise the pressure ratio, and therefore the polytropic head, in the first section in an attempt to equalize discharge temperatures. The first approximation can still be made by using Equation 5-19 or 5-20. However, the sectional pressure ratio will have to be modified by trial and error to obtain equalized discharge temperatures. This may take several iterations; one should be convinced that an acceptable temperature profile is obtained before proceeding too far into the compressor estimating procedure.

EXAMPLE 5-4

Consider a chlorine process for which the following average gas mixture properties are given:

$MW = 68$
$k = 1.3$
$Z = 0.95$

The inlet temperature is −32°C, the inlet pressure is 83 kPa. The object is to compress the gas to 600 kPa. The intercool heat exchanger has a pressure drop of 2% and is capable of a gas reentry temperature of 38°C. Process considerations indicate that the maximum allowable compression temperature is 115°C to minimize fouling. Determine the optimum pressure profile. Assume a polytropic efficiency of 77%.

First, assume two sections and estimate the pressure profile based on equal pressure ratios:

$$r_p = \frac{P_2}{P_1} = \frac{600}{83} = 7.23$$

$$r_{p\text{SECT}} = \sqrt{7.23} = 2.69$$

$$\frac{n}{n-1} = \frac{k}{k-1}\eta_p = \frac{1.3}{0.3}(0.77) = 3.34$$

Next consider the final discharge temperature for $r_{p\text{SECT}} = 2.69$, since the second-section discharge temperature will be higher than the first:

$$T_2 = T_1'(r_p)_2^{(n-1)/n} = 311\ (2.69)^{1/3.34} = 418\text{ K} = 145°\text{C}$$

The discharge temperature is too high. Arbitrarily reduce the second-section pressure ratio to 2.0 and try again:

$$T_2 = T_1'(r_p)_2^{(n-1)/n} = 311\ (2.0)^{1/3.34} = 383\text{ K} = 110°\text{C}$$

This is within process limitations.

Now review the pressure profile developed so far:

$$P_2 = 600\text{ kPa}$$

$$P_1' = \frac{P_2}{(r_p)_2} = \frac{600}{2} = 300\text{ kPa}$$

$$P_2' = 1.02\,P_1' = (1.02)(300) = 306\text{ kPa}$$

$$P_1 = 83\text{ kPa}$$

Thus:

$$(r_p)_1 = \frac{P_2'}{P_1} = \frac{306}{83} = 3.69$$

$$(r_p)_2 = 2.0$$

Calculate the first-section discharge temperature:

$$T_2' = T_1(r_p)_1^{(n-1)/n} = 241\ (3.69)^{1/3.34} = 356\text{ K} = 83°\text{C}$$

The first-section discharge temperature is lower than the second-section discharge temperature and further iteration is required.

It can be seen that the final discharge temperature was lowered by 35°C by reducing the pressure from 2.69 to 2.0. Also, there is a 27°C difference between the two sectional discharge temperatures according to the last iteration. It would appear that reducing the pressure ratio of the second section to approximately 1.75 would equalize the discharge temperatures. Using this new value, reestablish the pressure profile:

$$P_2 = 600 \text{ kPa}$$

$$P_1' = \frac{P_2}{(r_p)_2} = \frac{600}{1.75} = 343 \text{ kPa}$$

$$P_2' = 1.02\, P_1' = (1.02)(343) = 350 \text{ kPa}$$

$$P_1 = 83 \text{ kPa}$$

$$(r_p)_1 = \frac{P_2'}{P_1} = \frac{350}{83} = 4.22$$

$$(r_p)_2 = 1.75$$

$$T_2' = T_1(r_p)^{(n-1)/n} = 241\ (4.22)^{1/3.34} = 371 \text{ K} = 98°\text{C}$$

$$T_2 = T_1'(r_p)^{(n-1)/n} = 311\ (1.75)^{1/3.34} = 368 \text{ K} = 95°\text{C}$$

This would be an optimum pressure profile. If the required flow rate was in hand, the compressor selection could be continued as before.

The Multi-Body Compressor Train

The multi-body compressor train is one for which the polytropic head requirement is so large that the number of compression sections is too great to fit in a single compressor body. Since such cases will require one or more cool points due to the heat of compression, the solution to this problem parallels that developed for the intercooled compressor.

There is, however, one additional consideration when sizing the multi-body compressor train. When a gas is compressed, its volume decreases; thus the succeeding compressors will most likely be smaller frame sizes than the first. Smaller frames must operate at higher speeds in order to provide optimum duty. When two or more compressor bodies are connected together with the intention of using a common driver, it is not normally desirable to incorporate a gear between them. Gears add to train length and complicate the system rotor dynamics. Gears also add to maintenance requirements and are normally considered the weakest link in the chain. Depending on the driver selected, a three-body compressor train could conceivably require as many as three gears if each compressor were run at its optimum speed. If possible, a train would be best constructed with no gearing, but certainly one gear should be considered the practical limit.

How, then, will the combination of different frame size compressors and the lack of gearing complicate the compressor selection? Suppose that, based on process requirements, a Frame D compressor was to be directly connected to a Frame B compressor without gearing. The Frame D compressor operates nominally at 4,900 RPM, while the Frame B compressor operates nominally at 7,700 RPM. Suppose the head requirement from the Frame B compressor was 43,000 ft-lbf/lbm (129 k·Nm/kg). If this compressor operated at 7,700 RPM, it would take 5 impellers to perform the duty (43,000/10,000 or 129/30). However, when it is connected directly to Frame D, it only runs at 4,900 RPM; by the fan law, the polytropic head becomes:

$$H_{p_{4,900}} = H_{p_{\text{nom}}} \left(\frac{N}{N_{\text{nom}}}\right)^2 = 10{,}000 \left(\frac{4{,}900}{7{,}700}\right)^2 = 4{,}050 \text{ ft-lbf/lbm}$$

Thus, it would take 43,000/4,050 = 11 impellers to perform the duty, more than can normally be installed in a single compressor body. Also note that even if 11 impellers can be installed, the rotor length would be approximately twice as long as required at frame nominal speed. The speed imposed by the larger frame has considerably reduced the potential head production of the smaller frame. The result is that more compressor stages will be required.

One way to help alleviate the problem is to consider a double-flow compressor for the first (low-pressure) body. A double-flow compressor is one in which the flow is split equally between two inlet nozzles, one at each end of the compressor. The discharge nozzle is located in the center of the compressor. The double-flow arrangement essentially doubles the capacity of the compressor at the expense of approximately one-half the potential polytropic head production. The advantage to this arrangement is a smaller compressor frame and therefore better speed matching with downstream compressor bodies.

Consider the previous case, but this time assume the first body to be a double-flow Frame C instead of the single-flow Frame D. The Frame C compressor operates nominally at 5,900 RPM. How many stages will now be required in the Frame D compressor?

$$H_{p_{5,900}} = H_{p_{\text{nom}}} \left(\frac{N}{N_{\text{nom}}}\right)^2 = 10{,}000 \left(\frac{5{,}900}{7{,}700}\right)^2 = 5{,}870 \text{ ft-lbf/lbm}$$

$$\text{no. stages} = \frac{43{,}000}{5{,}870} = 8$$

The result is a satisfactory rotor length. Note that the resultant rotor will still be considerably longer than the five-impeller rotor required at Frame B nominal speed.

Now consider, by example, how to select a multi-body compressor train, the speed of the train, and the number of impellers required in each section of each compressor. The general approach is to follow the *N*-method procedure discussed earlier in this chapter; however, some additional steps are required for multi-body trains.

EXAMPLE 5-5

Consider a process requirement to compress 7,000 lbm/min of a charge gas stream from an inlet pressure of 16 psia to a discharge pressure of 515 psia. The gas stream is precooled and dried prior to entering the compressor train. The precooler and all intercool heat exchangers use water as the cooling medium. Each exchanger is capable of cooling the gas to 100°F with a 2% pressure drop. Process requirements further call for a cleanup duty external to the compressor before the gas enters the last compression section. The estimated pressure drop for this cleanup section and the associated intercool heat exchanger, piping, and fittings is 15 psi. Determine the number of compressor sections, number of compressor bodies, compressor frame sizes, number of stages in each section, string rotational speed (assuming direct drive, i.e. no gearing), total power requirement, and the discharge temperature for each section.

In actual practice, this example can be extremely tedious to estimate. The charge gas spoken of can contain as many as 20–30 measurable constituents. The mixture must first be analyzed to determine its inlet molecular weight. After estimating a discharge temperature for the first section, a *k*-value can be determined which in turn can be used to estimate the pressure profile. The charge gas will contain a significant percentage of high-molecular-weight gases such as propane and the butanes, pentanes, hexanes, and heptanes. Some of these heavier gases will tend to liquefy in one of the heat exchangers. The result is that the gas leaving the exchanger will have a lower molecular weight than that entering the exchanger. Also, the mass flow of gas leaving the cooler will be less than that entering, by the amount liquefied. The development of the gas properties throughout the compression process and the determination of the liquid knockout in the intercoolers are beyond the scope of this book; therefore they must be calculated by others means. To develop the compressor selection procedure for this case and to solve Example 5-5, the following assumptions are in order:

1. Assume that the molecular weight of the gas is 30.0 and that there is no liquid knockout in the intercoolers. Thus, all sections will have the same molecular weight and the same mass flow.

2. Assume that the average k-value is constant for all compressor sections and equal to 1.17. (Since all inlet temperatures are the same and since attempts are made to equalize the discharge temperatures, if the molecular weight in actual practice does not vary too greatly, the assumption of a constant average k-value throughout compression is valid.)
3. An assumption on average compressibility will be made after determining the number of compressor sections required. However, for simplicity, the average Z-value will be used to calculate the inlet flow to that section. This approach is certainly not entirely correct, but will serve as a good first approximation to establish the pressure profile, after which the gas analysis can be reevaluated for further compressor selection refinement.

As a first guess, assume that the compression process will require three sections, and that $\eta_p = 77\%$:

$$\frac{n}{n-1} = \frac{k}{k-1}\eta_p = \frac{1.17}{0.17}(0.77) = 5.30$$

$$r_p = \frac{P_2}{P_1} = 32.2$$

$$r_{p\,\mathrm{SECT}} = \sqrt[3]{r_p} = \sqrt[3]{32.2} = 3.18$$

$$T_2 = T_1(r_{p\,\mathrm{SECT}})^{(n-1)/n} = 560\,(3.18)^{1/5.30} = 697°\mathrm{R} = 237°\mathrm{F}$$

This is an acceptable discharge temperature. If the polytropic head requirement for each section is not too large (remember the limit on the practical number of impellers for each body), a three-section train can be selected.

Assume for now that a three-section compressor train will be acceptable. It was stated in the problem that the process required cleanup before entering the last compression section. This last section should be in a separate body, if possible, as the clean gas could then be totally isolated from the dirty gas. Note that if gas is removed from an intercooled compressor, cleaned, and reinjected into the same compressor, there would be the risk of dirty gas contaminating clean gas due to the cleanup pressure drop. Refer to Figure 5-5 for the schematic representation.

Since it is desirable to have the last section in a separate body, the first two sections will have to be placed in a single body to limit the compressor train length. Thus, for this example, the double-flow compressor cannot be used.

Now establish the pressure profile based on a three-section train and estimate the polytropic head requirements for each section in an effort to determine whether the process duty can be obtained with three compressor sections. Refer to Figure 5-6 for the nomenclature.

$$P_1 = 16 \text{ psia}$$

$$P_2' = P_1(r_{p\,\mathrm{SECT}})(1.02) = 16(3.18)(1.02) = 52 \text{ psia}$$

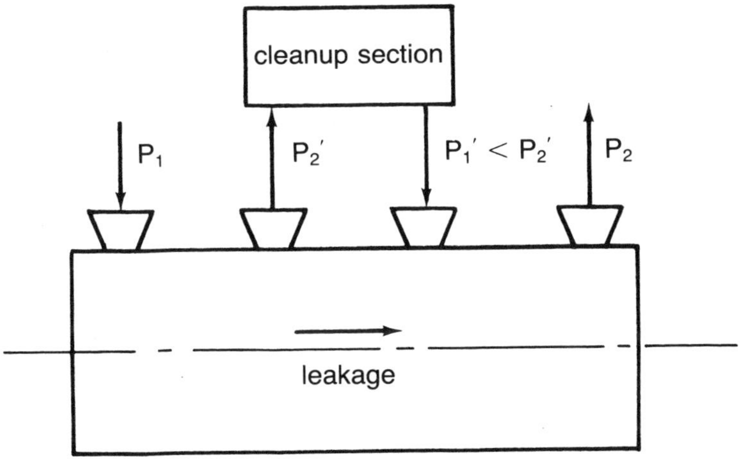

Figure 5-5. Intercooled compressor with cleanup section. Note that since the reentry pressure is lower than the intercool discharge pressure, there is a potential for dirty gas to leak into the clean-gas section.

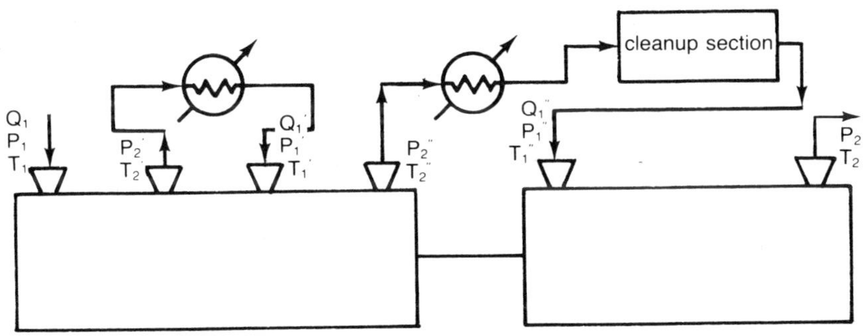

Figure 5-6. Three-section compressor train in two compressor bodies for Example 5-5.

$$P_1' = \frac{P_2'}{1.02} = \frac{52}{1.02} = 51 \text{ psia}$$

$$P_2'' = P_1' (r_{p\text{SECT}}) + 15 \text{ psia} = 51(3.18) + 15 = 177 \text{ psia}$$

$$P_1'' = P_2'' - 15 = 177 - 15 = 162 \text{ psia}$$

$$P_2 = 515 \text{ psia}$$

With this pressure profile, the average compressibility can be estimated by using either Figure C-5 or the compressibility charts in Appendix C. The parameters estimated so far are listed in Table 5-4.

Table 5-4
Estimated Parameters for Three-Section Compressor Train in Example 5-5

Section	1	2	3
Inlet pressure (psia)	16	51	162
Discharge pressure (psia)	52	177	515
Inlet temperature (°F)	100	100	100
k-value	1.17	1.17	1.17
$\frac{n}{n-1}$ value	5.30	5.30	5.30
Estimated Z-value	0.99	0.965	0.87
Molecular weight	30	30	30
Pressure ratio	3.25	3.47	3.18

$$(H_p)_1 = ZRT_1 \frac{n}{n-1} [r_p^{(n-1)/n} - 1] = (0.99)\left(\frac{1{,}545}{30}\right)(560)(5.30)[(3.25)^{1/5.30} - 1]$$

$$= 37{,}700 \text{ ft-lbf/lbm}$$

$$(H_p)_2 = (0.965)\left(\frac{1{,}545}{30}\right)(560)(5.3)[(3.47)^{1/5.30} - 1] = 39{,}000 \text{ ft-lbf/lbm}$$

$$(H_p)_3 = (0.87)\left(\frac{1{,}545}{30}\right)(560)(5.3)[(3.18)^{1/5.30} - 1] = 32{,}400 \text{ ft-lbf/lbm}$$

Recall the foregoing discussion where a Frame B compressor was directly connected to a Frame D. It was illustrated that by jumping two frame sizes the polytropic head was reduced to approximately 40% of the nominal. Also, it has been assumed that nine impellers is the maximum practical limit for a frame size for estimating purposes. Therefore, a two-frame jump would imply that the polytropic head required from the smaller body should not exceed approximately 9 × 0.4 × 10,000, or 36,000 ft-lbf/lbm (108 k·Nm/kg). Actual limits will be set by the manufacturer, since he may have at his disposal higher head impellers to boost the head production at the lower speeds. Since the volume flow into the second body will be on the order of 10% of the volume flow into the first body (16/150), it may be necessary to jump three frame sizes for the second body. This would even further reduce the maximum head production.

This case exceeds the 36,000 ft-lbf/lbm guideline. While a three-section compressor train might prove feasible, a three-section train has a slim chance of success. Based on this, one should consider a four-section compressor and proceed with the estimate.

With a four-section compressor, it is still beneficial to isolate the last section after cleanup in a separate body. This leaves three sections, which is too many for one body. Thus, a three-body compressor train results. Perhaps the best way to split the remaining three sections would be to use a double-flow compressor

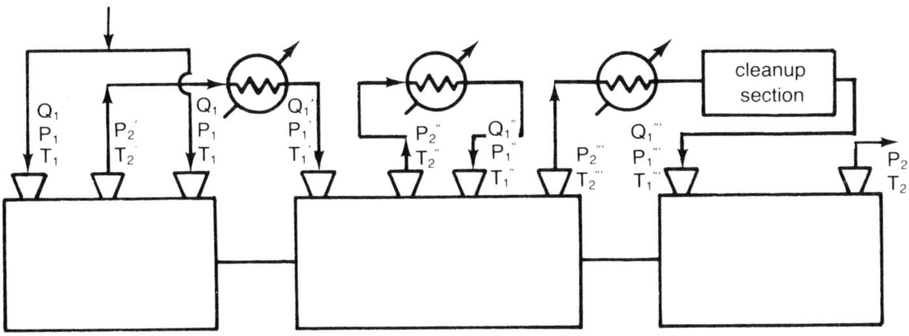

Figure 5-7. Four-section compressor train in three compressor bodies for Example 5-5.

followed by an intercooled compressor. This approach would permit one to take advantage of the better speed matching afforded by the double-flow arrangement. This compressor train is schematically represented in Figure 5-7.

Now reestablish the pressure profile:

$$r_p = \frac{P_2}{P_1} = \frac{515}{16} = 32.2$$

$$r_{p\text{SECT}} = \sqrt[4]{r_p} = \sqrt[4]{32.2} = 2.38$$

$$P = 16 \text{ psia}$$

$$P_2' = P_1(r_{p\text{SECT}})(1.02) = 16(2.38)(1.02) = 39 \text{ psia}$$

$$P_1' = \frac{P_2'}{1.02} = \frac{39}{1.02} = 38.2 \text{ psia}$$

$$P_2'' = P_1'(r_{p\text{SECT}})(1.02) = 38.2(2.38)(1.02) = 93 \text{ psia}$$

$$P_1'' = \frac{P_2''}{1.02} = \frac{93}{1.02} = 91.2 \text{ psia}$$

$$P_2''' = P_1''(r_{p\text{SECT}}) + 15 \text{ psia} = 91.2(2.38) + 15 = 232 \text{ psia}$$

$$P_1''' = P_2''' - 15 \text{ psia} = 232 - 15 = 217 \text{ psia}$$

$$P_2 = 515 \text{ psia}$$

The resulting pressure profile and other known values are reflected in Table 5-5.

Look at the first section. Remember that for simplicity the average compressibility will be used for inlet Z-value.

$$Q_1 = m_1 \frac{Z_1 R T_1}{144 P_1} = (7{,}000)\frac{(0.99)(1{,}545)(560)}{(30)(144)(16)} = 86{,}745 \text{ ICFM}$$

Table 5-5
Estimated Parameters for Four-Section Compressor Train in Example 5-5

Section	1	2	3	4
Inlet pressure (psia)	16	38.2	91.2	217
Discharge pressure (psia)	39	93	232	515
Inlet temperature (°F)	100	100	100	100
k-value	1.17	1.17	1.17	1.17
Average *Z*-value	0.99	0.98	0.95	0.86
Molecular weight	30	30	30	30
Pressure ratio	2.44	2.43	2.54	2.37

Since the selection will be a double-flow compressor for the first body, a frame size must be selected for half the volume flow, or 43,372 ICFM. Select a Frame D compressor. While the flow rate is very close to the maximum nominal flow for this frame, the compressor will be less expensive than a Frame E and will provide a better speed match with the smaller compressors downstream.

$$\eta_p = 77\% \quad \text{from Table 5-1}$$

$$\frac{n}{n-1} = \frac{k}{k-1}\eta_p = \frac{1.17}{0.17}(0.77) = 5.30$$

$$(H_p)_1 = Z_a R T_1 \frac{n}{n-1}\left[(r_p)_1^{(n-1)/n} - 1\right]$$

$$= (0.99)\left(\frac{1{,}545}{30}\right)(560)(5.30)[(2.44)^{1/5.30} - 1] = 27{,}700 \text{ ft-lbf/lbm}$$

$$T_2' = T_1 (r_p)_1^{(n-1)/n} = 560\ (2.44)^{1/5.30} = 663°\text{R} = 203°\text{F}$$

Determine the number of stages:

$$\theta = \sqrt{\frac{26.1\, MW}{k_1 Z_1 T_1}} = \sqrt{\frac{(26.1)(30)}{(1.17)(0.99)(560)}} = 1.10$$

$$\text{max. } H_p/\text{stage} = 12{,}000 \text{ ft-lbf/lbm}$$

$$\text{no. stages} = \frac{H_p}{\text{max. } H_p/\text{stage}} = \frac{27{,}700}{12{,}000} = 3$$

Now determine the speed:

$$N = N_{nom}\sqrt{\frac{H_p}{H_{p_{nom}} \times \text{no. stages}}} = 4{,}900\sqrt{\frac{27{,}700}{(10{,}000)(3)}} = 4{,}710 \text{ RPM}$$

(In order to keep the speed as high as possible for the benefit of the downstream compressors, one may wish to consider the first body maximum head to be 10,000 ft-lbf/lbm. In this example, three impellers would still have been calculated for the first section.)

The speed of the train is now fixed at 4,710 RPM. An attempt will be made to select two compressors to round out the train that will provide the necessary duty at 4,710 RPM. Look at the second section:

$$Q_1' = m_1' \frac{Z_1' R T_1'}{144\, P_1'} = (7{,}000) \frac{(0.98)(1{,}545)(560)}{(30)(144)(38.2)} = 35{,}966 \text{ ICFM}$$

This, too, is a Frame D compressor with a polytropic efficiency of 77%.

$$(H_p)_2 = Z_a' R\, T_1' \frac{n}{n-1} [(r_p)_2^{(n-1)/n} - 1]$$

$$= (0.98)\left(\frac{1{,}545}{30}\right)(560)(5.30)[(2.43)^{1/5.30} - 1]$$

$$= 27{,}300 \text{ ft-lbf/lbm}$$

$$T_2'' = T_1' (r_p)_2^{(n-1)/n}$$

$$= 560(2.43)^{1/5.30} = 662°\text{R} = 202°\text{F}$$

Now compute the number of impellers for this section; the fixed speed is already known. Consider Equation 5-15:

$$N = N_{\text{nom}} \sqrt{\frac{H_p}{H_{p_{\text{nom}}} \times \text{no. stages}}} \qquad (5\text{-}15)$$

which can be rearranged as follows:

$$\text{no. stages} = \left(\frac{N_{\text{nom}}}{N}\right)^2 \left(\frac{H_p}{H_{p_{\text{nom}}}}\right) \qquad (5\text{-}21)$$

Then the number of stages will be:

$$\text{no. stages} = \left(\frac{N_{\text{nom}}}{N}\right)^2 \left(\frac{H_p}{H_{p_{\text{nom}}}}\right) = \left(\frac{4{,}900}{4{,}710}\right)^2 \left(\frac{27{,}300}{10{,}000}\right) = 2.95 \rightarrow 3$$

Note that θ was not checked, since the values of θ are nearly equal to those of the first body.

Now the third section, regardless of flow, is a Frame D, since this is the second section of the intercooled compressor:

$$(H_p)_3 = Z''_a R T''_1 \frac{n}{n-1} [(r_p)_3^{(n-1)/n} - 1]$$

$$= (0.95)\left(\frac{1{,}545}{30}\right)(560)(5.30)[(2.54)^{1/5.30} - 1]$$

$$= 27{,}900 \text{ ft-lbf/lbm}$$

$$T_2''' = T_1''(r_p)_3^{(n-1)/n} = 560\ (2.54)^{1/5.30} = 668°\text{R} = 208°\text{F}$$

$$\text{no. stages} = \left(\frac{N_{\text{nom}}}{N}\right)^2 \left(\frac{H_p}{H_{p_{\text{nom}}}}\right) = \left(\frac{4{,}900}{4{,}710}\right)^2 \left(\frac{27{,}900}{10{,}000}\right) = 3.02 \rightarrow 3$$

Three impellers are required instead of four, since the calculated value is so close to three that one can make up in other sections, by a speed increase if necessary, the little lost in the third section. Only very minor pressure profile variations will result.

Now consider the fourth section:

$$Q_1''' = m_1''' \frac{Z_1''' R T_1'''}{144\ P_1'''} = (7{,}000) \frac{(0.86)(1{,}545)(560)}{(30)(144)(217)}$$

$$= 5{,}556 \text{ ICFM}$$

Which frame should be selected? At 4,710 RPM the maximum flow for Frame A will be reduced by the ratio of 4,710/11,000 as implied by the fan law. Therefore, the maximum flow for Frame A becomes:

$$\text{max. flow}_A = \left(\frac{4{,}710}{11{,}000}\right)(7{,}000) = 3{,}000 \text{ ICFM}$$

which is too low. Check Frame B:

$$\text{max. flow}_B = \left(\frac{4{,}710}{7{,}700}\right)(18{,}000) = 11{,}000 \text{ ICFM}$$

$$\text{min. flow}_B = \left(\frac{4{,}710}{7{,}700}\right)(6{,}000) = 3{,}700 \text{ ICFM}$$

Therefore the required flow falls in the range of Frame B. The polytropic efficiency is 76%. θ need not be checked, since at such a low speed the maximum polytropic head per stage for Frame B cannot be exceeded.

$$\frac{n}{n-1} = \frac{k}{k-1}\eta_p = \frac{1.17}{0.17}(0.76) = 5.23$$

$$(H_p)_4 = Z_a'''RT_1''' \frac{n}{n-1} [(r_p)_4^{(n-1)/n} - 1]$$

$$= (0.86)\left(\frac{1,545}{30}\right)(560)(5.23)[(2.37)^{1/5.23} - 1] = 23,300 \text{ ft-lbf/lbm}$$

$$T_2 = T_1'''(r_p)_4^{(n-1)/n} = 560(2.37)^{1/5.23} = 660°\text{R} = 200°\text{F}$$

$$\text{no. stages} = \left(\frac{N_{\text{nom}}}{N}\right)^2 \left(\frac{H_p}{H_{p_{\text{nom}}}}\right) = \left(\frac{7,700}{4,710}\right)^2 \left(\frac{23,300}{10,000}\right) = 6.23 \rightarrow 7$$

The power requirement for the train will be the sum of the power requirements for each compressor:

First body:

$$(PWR_g)_1 = \frac{m_1(H_p)_1}{33,000(\eta_p)_1} = \frac{(7,000)(27,700)}{(33,000)(0.77)} = 7,630 \text{ hp}$$

$$(L_m)_1 = (.02)(7,630) = 153 \text{ hp}$$

Second body:

$$(PWR_g)_2 = \frac{m_1'(H_p)_2}{33,000\,(\eta_p)_2} + \frac{m_1''(H_p)_3}{33,000\,(\eta_p)_3}$$

$$= \frac{(7,000)(27,300)}{(33,000)(0.77)} + \frac{(7,000)(27,900)}{(33,000)(0.77)} = 15,210 \text{ hp}$$

$$(L_m)_2 = (.015)(15,210) = 228 \text{ hp}$$

Third body:

$$(PWR_g)_3 = \frac{m_1'''(H_p)_4}{33,000\,(\eta_p)_4} = \frac{(7,000)(23,300)}{(33,000)(0.76)} = 6,505 \text{ hp}$$

$$(L_m)_3 = (.02)(6505) = 130 \text{ hp}$$

Total power:

$$PWR_s = (PWR_g)_1 + (PWR_g)_2 + (PWR_g)_3 + (L_m)_1 + (L_m)_2 + (L_m)_3$$

$$= 7,630 + 15,210 + 6,505 + 153 + 228 + 130$$

$$\approx 29,900 \text{ hp}$$

Table 5-6
Solution to Example 5-5

Section	1	2	3	4
Mass flow (lbm/min)	7,000	7,000	7,000	7,000
Inlet pressure (psia)	16	38.2	91.2	217
Discharge pressure (psia)	39	93	232	515
Inlet temperature (°F)	100	100	100	100
Discharge temperature (°F)	203	202	208	200
Inlet volume flow (ICFM)	86,745	35,966	—	5,556
Polytropic head (ft-lbf/lbm)	27,700	27,300	27,900	23,300
Number of stages	3	3	3	7

Rotational speed (RPM): 4,710
Power requirement (hp): 29,900
Number of compressor bodies: 3

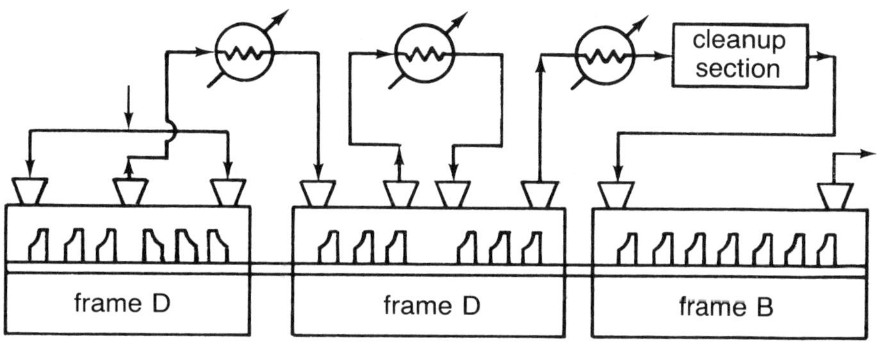

Figure 5-8. Schematic representation of the solution to Example 5-5.

Thus, the selection is a three-body direct-connected compressor train: a double-flow Frame D compressor coupled to an intercooled Frame D compressor coupled to a straight-through Frame B compressor. Its performance is tabulated in Table 5-6 and schematically represented in Figure 5-8.

The Side-Load Compressor

A side-load compressor is one in which gas is either injected into or extracted from the compressor casing somewhere between the main inlet and discharge nozzles. The side-load compressor is primarily used in refrigeration processes where two or more pressure levels are required for process refrigeration. Schematically, the side-load compressor is as illustrated in Figure 5-9.

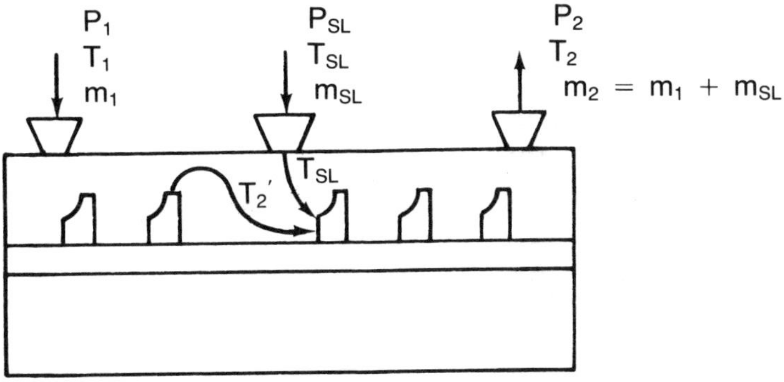

Figure 5-9. Schematic representation of a side-load compressor.

The *N*-method can be used to estimate the sizing of a side-load compressor similar to the way the intercooled compressor is estimated. With a side-load, however, there is one additional consideration: The mixture temperature entering the first wheel of the section downstream of the side-load nozzle must be determined. The mixture temperature then becomes the inlet temperature to the next section for the calculation of volume flow and polytropic head.

Temperatures mix based on the relative mass flows and specific heats of each gas stream; i.e. one really mixes enthalpies. The correct way to mix the gas stream temperatures is by mass weighting enthalpies:

$$T_{mix} = \frac{m_1(c'_{p_2})(T'_2) + m_{SL}(c_{p_{SL}})(T_{SL})}{m_1(c'_{p_2}) + m_{SL}(c_{p_{SL}})} \tag{5-22}$$

Note that Equation 5-22 is valid when temperature is in °F, °C, °R, or K.

For quick estimates, mixture temperature can be approximated by a weighted average of the stream temperatures without knowing c_p:

$$T_{mix} \approx \frac{m_1 T'_2 + m_{SL} T_{SL}}{m_1 + m_{SL}} \tag{5-23}$$

Equation 5-23 is not as accurate as Equation 5-22; it should only be used for rough calculations.

A mixture temperature must be calculated for each side-load considered.

While the *N*-method is a means of compressor sizing for side-load units, a more accurate approach is the use of a Mollier chart. When one exists for the gas under consideration, the compressor should be selected by the Mollier method.

Mollier Method for Estimating Compressor Sizing

The second estimating procedure to investigate is the Mollier method. This procedure is more accurate than the *N*-method; it should be employed when working with a pure gas for which a Mollier chart is available (and even a gas mixture if a Mollier chart is made available). Perhaps a good definition for gas purity would be 95+% of a single constituent. In practice, the Mollier method is most often used for side-load or extraction applications. The Mollier method is as follows (refer to Figure 5-10 for state-point nomenclature):

1. After plotting the inlet point knowing P_1 and T_1, determine the inlet specific volume, v_1, from the Mollier chart.
2. Determine the inlet volume flow from Equation 4-1:

$$Q_1 = m_1 v_1 \tag{4-1}$$

3. Select the compressor frame by using the calculated inlet volume flow and Table 5-1.
4. Determine the adiabatic head:
   a. Read the inlet enthalpy, h_1, directly from point ① on the Mollier chart.
   b. Follow a constant entropy line to the required discharge pressure, P_2, and plot the point. This point, (2ad), is the adiabatic isentropic discharge state point for the first section of the compressor.
   c. Read the adiabatic discharge enthalpy, h_{2ad}, for state point (2ad).
   d. Calculate the adiabatic head from Equation 5-24 or 5-25:

$$H_{ad} = 778\,(h_{2ad} - h_1) \quad \text{in English system} \tag{5-24}$$

$$H_{ad} = h_{2ad} - h_1 \quad \text{in metric system} \tag{5-25}$$

5. Determine the polytropic head:
   a. Determine the *k*-value from Table C-1 in Appendix C. A little more accuracy may be obtained by calculating the *k*-value at the actual temperature; however, the tabulated value will normally provide sufficient accuracy.
   b. Determine the section pressure ratio, r_p.
   c. Determine the adiabatic efficiency from Figure 5-11.
   d. Determine the polytropic head from Equation 5-26:

$$H_p = H_{ad}\left(\frac{\eta_p}{\eta_{ad}}\right) \tag{5-26}$$

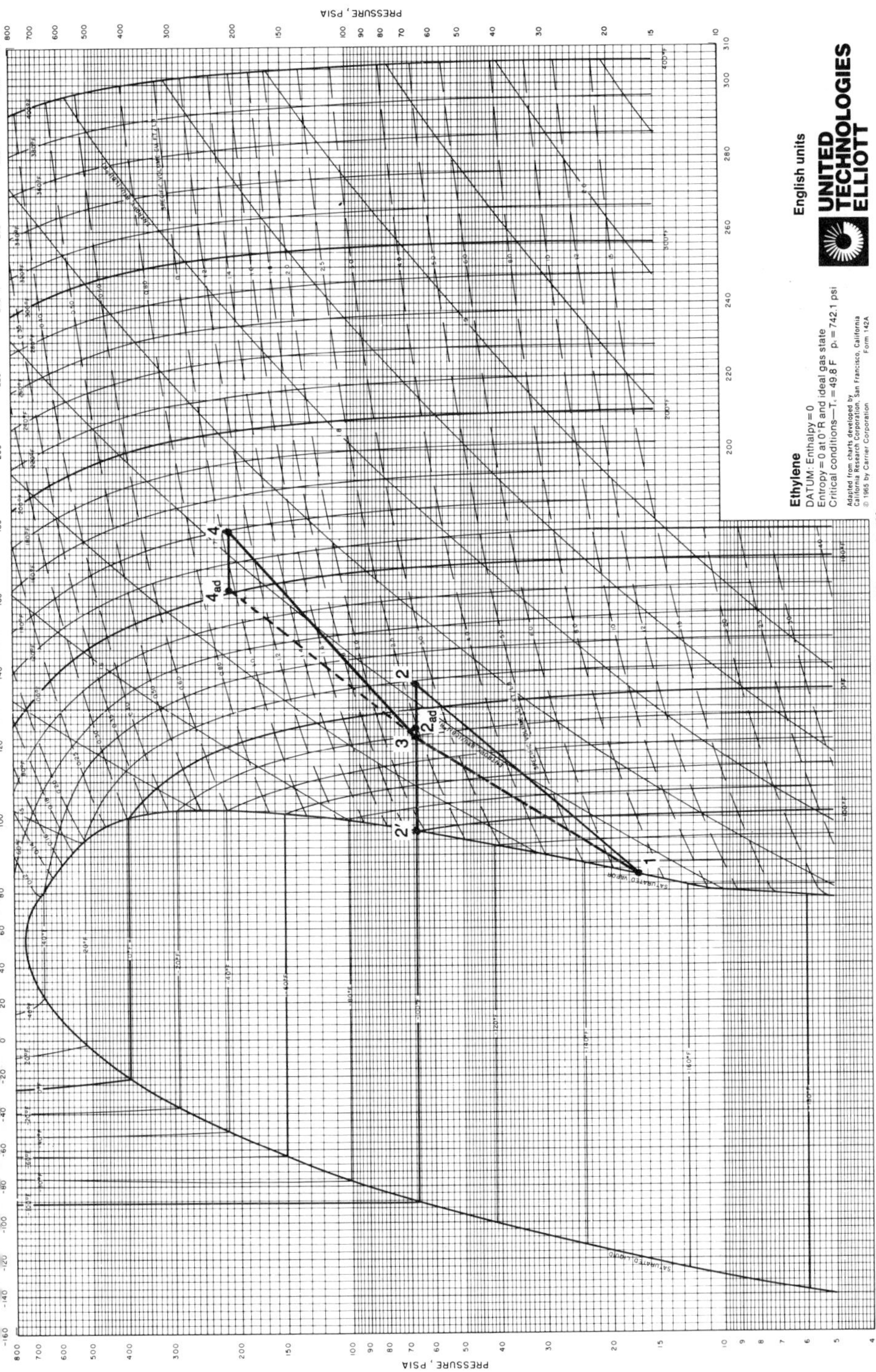

Figure 5-10. Ethylene Mollier chart for Example 5-6.

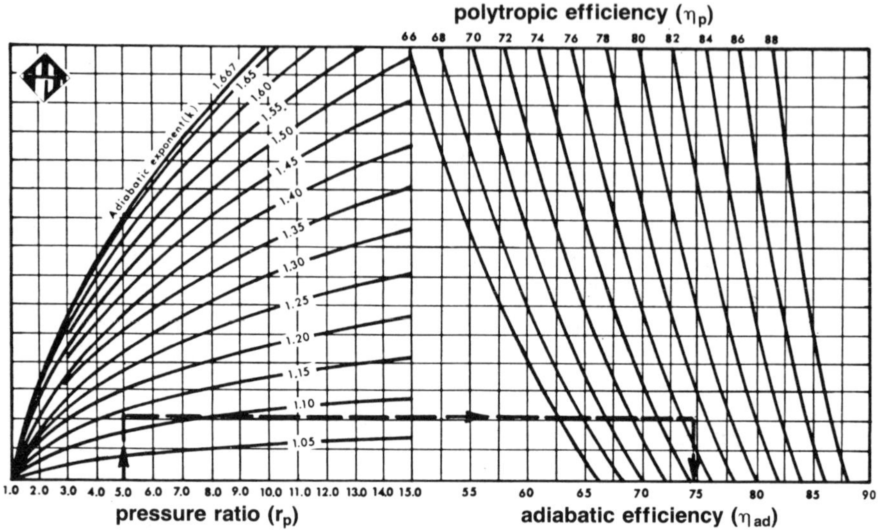

Figure 5-11. Efficiency conversion chart. Reprinted by permission of Gas Processors Suppliers Association *Engineering Data Book,* Ninth Edition, 1972, 5th Revision, 1981.

6. From Equation 5-27, determine the actual discharge state point for the first section and plot this as state-point ②:

$$h_2 = \frac{h_{2ad} - h_1}{\eta_{ad}} + h_1 \tag{5-27}$$

Also determine T_2 directly from the Mollier chart for state-point ②.

7. Determine state-point ③, which is the inlet state point to the second section of compression:
   a. Plot state-point ②', the side-load level, based on known side-load pressure and temperature.
   b. Read the enthalpy for state-point ②' directly from the Mollier chart.
   c. Determine the mixture enthalpy for the two gas streams by using Equation 5-28:

$$h_3 = \frac{m_1 h_2 + m_{SL} h_{2'}}{m_1 + m_{SL}} \tag{5-28}$$

where:

m_{SL} = side-load mass flow

Note also that the mass flow to the second section is the sum of the first section and side-load mass flows:

$$m_2 = m_1 + m_{SL} \tag{5-29}$$

8. Follow steps 4, 5, and 6 for the second section.
9. If there are two side-loads, repeat the foregoing to establish state points ⑤ and ⑥.
10. Determine the required number of compressor stages for each section:
    a. Determine the value of θ from Equation 5-30 or 5-31:

$$\theta = \sqrt{\frac{280}{k_1 P_1 v_1}} \quad \text{in English system} \tag{5-30}$$

$$\theta = \sqrt{\frac{121}{k_1 P_1 v_1}} \quad \text{in metric system} \tag{5-31}$$

    Note that Equations 5-30 and 5-31 are identical to Equations 5-12 and 5-13 except that Pv has been substituted for ZRT. Here, again, the k-values tabulated in Table C-1 can be used, since minor errors in θ will not affect the estimate.

    b. Determine the maximum polytropic head per stage from either Figure 5-2 or 5-3.
    c. From Equation 5-14, determine the number of stages for each section:

$$\text{no. stages} = \frac{H_p}{\text{max. } H_p/\text{stage}} \tag{5-14}$$

11. From Equation 5-15, determine the required rotational speed for each section and use the higher speed value:

$$N = N_{\text{nom}} \sqrt{\frac{H_p}{H_{p_{\text{nom}}} \times \text{no. stages}}} \tag{5-15}$$

12. Determine the required shaft power:
    a. Determine the gas power requirement for each section from Equation 5-16 or 5-17:

$$PWR_g = \frac{m H_p}{33{,}000\ \eta_p} \quad \text{in English system} \tag{5-16}$$

$$PWR_g = \frac{m H_p}{3{,}600\ \eta_p} \quad \text{in metric system} \tag{5-17}$$

    b. Determine the mechanical losses from Table 5-3.
    c. Determine the required shaft power from equation (5-18):

$$PWR_s = PWR_g + L_m \tag{5-18}$$

13. Determine the final discharge temperature directly from the Mollier chart.

EXAMPLE 5-6

A compressor is required to perform the following duty for a refrigeration service. It must compress 900 lbm/min of saturated ethylene from 17 psia to 213 psia, with a 500 lbm/min side-load at 66 psia, saturated. Determine the compressor frame size, number of stages, rotational speed, power requirement, and discharge temperature using the Mollier method. Refer to Figure 5-10.

1. Determine the inlet specific volume:

$$v_1 = 7.1\ \text{ft}^3/\text{lbm}$$

2. Determine the inlet volume flow:

$$Q_1 = m_1 v_1 = (900)(7.1) = 6{,}390\ \text{ICFM}$$

3. Select the compressor frame from Table 5-1. Select Frame A with a polytropic efficiency of 76%.

4. Determine the adiabatic head:

$$h_1 = 84\ \text{Btu/lbm}$$

$$h_{2ad} = 122\ \text{Btu/lbm}$$

$$H_{ad} = 778(h_{2ad} - h_1) = 778(122 - 84) = 29{,}600\ \text{ft-lbf/lbm}$$

5. Determine the polytropic head:

$$k = 1.24$$

$$r_p = \frac{P_2}{P_1} = \frac{66}{17} = 3.88$$

$\eta_{ad} = 72.5\%$ from Figure 5-11

$$H_p = H_{ad}\left(\frac{\eta_p}{\eta_{ad}}\right) = 29{,}600\left(\frac{0.760}{0.725}\right) = 31{,}000 \text{ ft-lbf/lbm}$$

6. Determine the actual discharge enthalpy:

$$h_2 = \frac{h_{2ad} - h_1}{\eta_{ad}} + h_1 = \frac{122 - 84}{0.725} + 84 = 136.4 \text{ Btu/lbm}$$

7. Determine state-point ③:

$h_{2'} = 96$ Btu/lbm from Mollier chart

$$h_3 = \frac{m_1 h_2 + m_{SL} h_{2'}}{m_1 + m_{SL}} = \frac{(900)(136.4) + (500)(96.0)}{900 + 500}$$

$$= 122 \text{ Btu/lbm}$$

$$m_2 = m_1 + m_{SL} = 900 + 500 = 1{,}400 \text{ lbm/min}$$

8. Follow steps 4, 5, and 6 for the second section:

$$h_3 = 122 \text{ Btu/lbm}$$

$$h_{4ad} = 162 \text{ Btu/lbm}$$

$$H_{ad} = 778(h_{4ad} - h_3) = 778(162 - 122) = 31{,}100 \text{ ft-lbf/lbm}$$

$$k = 1.24$$

$$r_p = \frac{P_4}{P_3} = \frac{213}{66} = 3.23$$

$$\eta_{ad} = 73\%$$

$$H_p = H_{ad}\left(\frac{\eta_p}{\eta_{ad}}\right) = 31{,}100\left(\frac{0.76}{0.73}\right) = 32{,}400 \text{ ft-lbf/lbm}$$

$$h_4 = \frac{h_{4ad} - h_3}{\eta_{ad}} + h_3 = \frac{162 - 122}{0.73} + 122 = 176.8 \text{ Btu/lbm}$$

9. There is only one side-load.
10. Determine the required number of compressor stages for each section:

$$\theta = \sqrt{\frac{280}{k_1 P_1 v_1}} = \sqrt{\frac{280}{(1.24)(17)(7.1)}} = 1.37$$

max. H_p/stage = 10,900 ft-lbf/lbm from Figure 5-2

$$\text{no. stages} = \frac{H_p}{\text{max. } H_p/\text{stage}}$$

Section 1

$$\text{no. stages} = \frac{31{,}000}{10{,}900} = 2.84 \rightarrow 3 \tag{5-14}$$

Section 2

$$\text{no. stages} = \frac{32{,}400}{10{,}900} = 2.97 \rightarrow 3$$

11. Determine the required rotational speed for each section:

$$N = N_{\text{nom}} \sqrt{\frac{H_p}{H_{p_{\text{nom}}} \times \text{no. stages}}}$$

Section 1

$$N = 11{,}000 \sqrt{\frac{31{,}000}{10{,}000 \times 3}} = 11{,}200 \text{ RPM} \tag{5-15}$$

Section 2

$$N = 11{,}000 \sqrt{\frac{32{,}400}{10{,}000 \times 3}} = 11{,}450 \text{ RPM}$$

Therefore, the speed will be 11,450 RPM.

12. Determine the required shaft power:

$$PWR_g = \frac{m H_p}{33{,}000\, \eta_p}$$

Section 1

$$PWR_g = \frac{(900)(31{,}000)}{(33{,}000)(0.76)} = 1{,}115 \text{ hp}$$

Section 2

$$PWR_g = \frac{(1{,}400)(32{,}400)}{(33{,}000)(0.76)} = 1{,}810 \text{ hp}$$

Total

$$PWR_g = 1{,}115 + 1{,}810 = 2{,}925 \text{ hp}$$

$$L_m = (.03)(2{,}925) = 88 \text{ hp}$$

$$PWR_s = PWR_g + L_m = 2{,}925 + 88 \approx 3{,}015 \text{ hp}$$

13. Determine the final discharge temperature directly from the Mollier chart:

$$T_4 = 138°\text{F}$$

Therefore, the selection is a Frame A compressor with three stages (impellers) in each section, operating at 11,450 RPM, requiring 3,015 hp, with a discharge temperature of 138°F.

A Final Test for Feasibility

Manufacturers control the head production from an impeller in two ways. The first is speed control, which has already been discussed extensively; the other is impeller diameter trimming, which will be looked at now.

A manufacturer can trim up to about 20% of the polytropic head out of an impeller by cutting back on the impeller blade diameter. Within this range, the impeller will normally lose only a small amount of efficiency. The current state of the art generally prohibits a manufacturer from trimming any more than 20% of the head. Therefore, each impeller is capable of 80%–100% polytropic head at a specified speed.

How does this affect estimates? Suppose one has a two-section compressor; the speed is fixed by some consideration independent of the present analysis. Process pressure levels, the resulting polytropic head, and the fixed speed indicate the need for 3.0 impellers in the first section and 2.1 impellers in the second section. Obviously, three impellers will be selected for each section. However, three impellers in the second section will produce too

Table 5-7
Minimum Calculated Number of Stages per Section

Actual Number of Stages	Minimum Calculated Number of Stages
1	0.8
2	1.6
3	2.4
4	3.2

much head. If the manufacturer trims 20% head out of each second-section impeller, the head production will still be too high, since 3 × 0.8 = 2.4 effective impellers would be had, while only 2.1 are required.

This phenomenon will not present problems with straight-through or intercooled compressors, or even with multi-body compressor trains. For these applications, the speed of the compressor can be slightly modified and pressure levels altered to provide sufficiently balanced sectional head requirements for feasible compressor impeller selection.

However, a feasibility problem could be witnessed with side-load compressors. Side-load compressors are normally used for refrigeration duty. For refrigeration compressors, side-load pressures are based on desired temperature levels for refrigeration. Usually, the manufacturer is only given minimal tolerance to work with on these pressure levels. A poorly set pressure profile may force the manufacturer into an inefficient impeller selection. It is even possible that the duty becomes impractical, if not impossible, to satisfy.

Consider a case where process requirements for a 3-section side-load compressor dictate sectional polytropic heads of 5,000, 20,000 and 30,000 ft-lbf/lbm, respectively. If compressor speed is based on the first-section duty, the speed for 5,000 ft-lbf/lbm/stage will be calculated. The result would be 11 stages, requiring 2 compressor bodies. If the pressure levels are shifted to redistribute the 55,000 ft-lbf/lbm more efficiently, say 10,000, 20,000 and 25,000 ft-lbf/lbm, a compressor could easily be sized for 10,000 ft-lbf/lbm/stage. The result is 6 stages and 1 compressor body. Note that while the 3 stages in the third section are capable of 30,000 ft-lbf/lbm as full-diameter impellers, each impeller could be trimmed to 80% head, or 24,000 ft-lbf/lbm, for the section. Therefore, any head requirement from 24,000 to 30,000 ft-lbf/lbm is practical from this section.

One should always bare this in mind when sizing side-load compressors. The calculated number of stages per section should never be less than those tabulated in Table 5-7.

6

The Centrifugal Compressor Performance Curve

Having reviewed the methods for estimating compressor sizing and power requirements, it is time to look at the compressor performance curve.

The Centrifugal Compressor Characteristic

Consider the simple model in Figure 6-1 which shows a compressor taking air at atmospheric pressure and discharging it into a receiver. The compressor and receiver are both initially at atmospheric pressure, since they are in equilibrium with their surroundings. The instant the compressor starts, the compressor inlet and discharge pressures will be equal because of the prior equilibrium. Since the discharge pressure equals the inlet pressure, there is essentially no resistance to flow (neglecting pipe friction), and the pressure rise from inlet to discharge is zero. As a result, the head produced by the compressor is zero (refer to the polytropic head equation, Equation 2-12). This point is plotted on Figure 6-2 as Point 1. As the mass of air in the receiver starts to increase, the pressure in the receiver will start to rise, providing some resistance. At first, the flow will drop very slightly. This point is shown on Figure 6-2 as Point 2; it is called the stonewall point, since the curve beyond this point is essentially a straight line.

As the mass of air in the receiver continues to increase, the pressure also increases, providing greater pressure differential from inlet to discharge of the compressor and, at the same time, a greater resistance to flow (it becomes more difficult to "cram" the air into the receiver). This is shown as Point 3 on Figure 6-3 and is a typical compressor operating point.

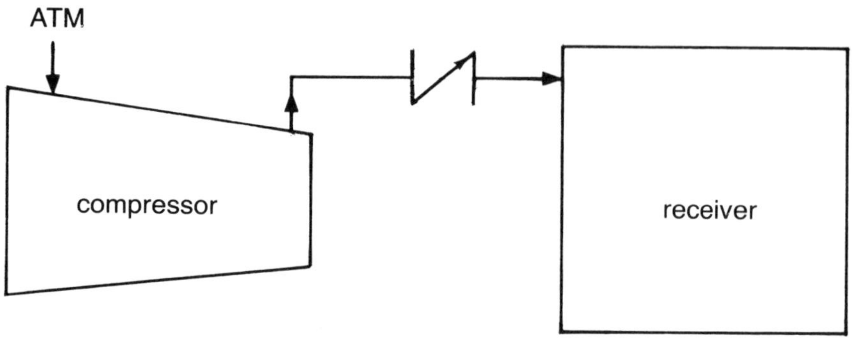

Figure 6-1. Compressor delivering air into a receiver.

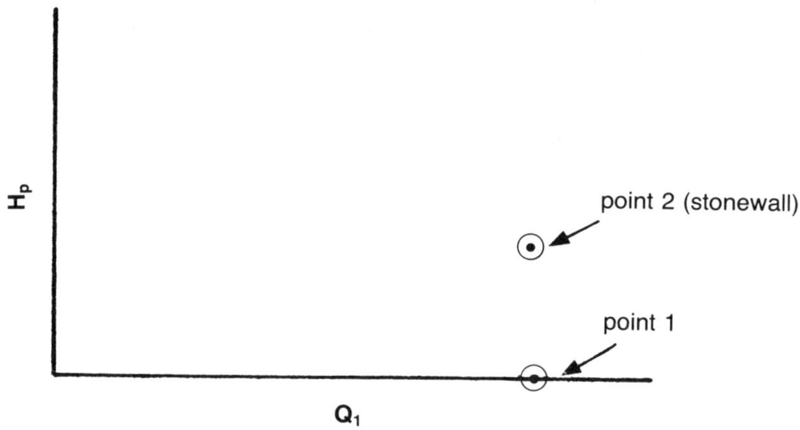

Figure 6-2. Compressor volume flow with no resistance—stonewall.

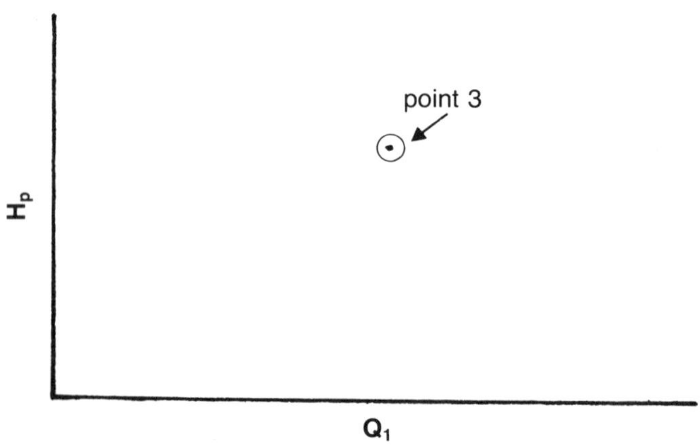

Figure 6-3. Compressor volume flow with some resistance.

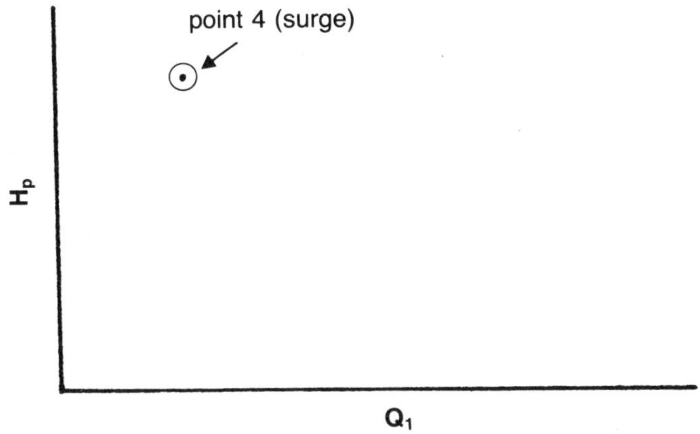

Figure 6-4. Compressor volume flow at maximum resistance—surge.

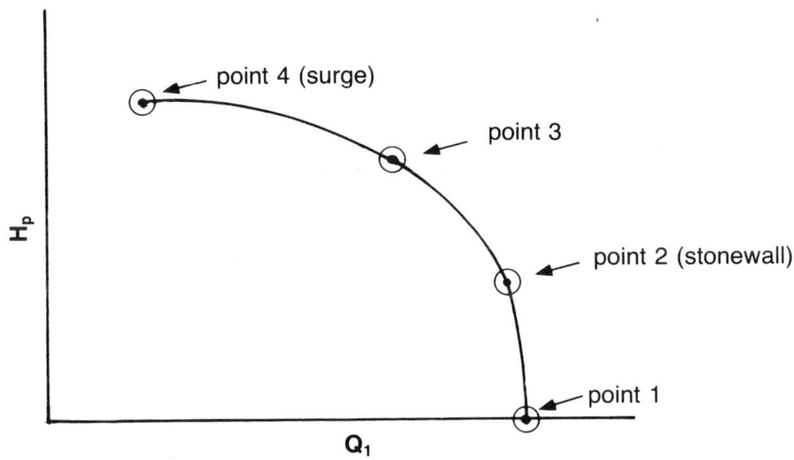

Figure 6-5. The centrifugal compressor characteristic curve.

As the mass of air in the receiver increases further, a pressure is eventually obtained, above which the compressor cannot pump stably. This point, shown as Point 4 on Figure 6-4, is called the surge point.

If all these points are plotted on a single curve and connected, the result is the centrifugal compressor characteristic curve shown in Figure 6-5.

In this analysis, the three important points are the stonewall point (Point 2), the operating point (Point 3), and the surge point (Point 4). Chapter 5 was devoted to estimating the operating point. What are the stonewall and surge points and what roles do they play?

The stonewall point is the maximum stable compressor flow point. Consequently, it is the minimum head point under stable compressor operation. Flow increases beyond stonewall are minimal at best; operation beyond the stonewall point is unpredictable because of the vertical slope of the curve in this area. Manufacturers will usually stop their curves before they reach the vertical section. They only extend the curve as far as they can reasonably predict performance.

The surge point is the minimum stable flow point; consequently, it is the highest head point. As the pressure in the receiver continues to increase, the volume flow rate into the receiver becomes less and less and therefore the velocity of the gas going through the compressor and into the piping continues to decrease. When the velocity becomes too slow, the compressor can no longer perform stably. Since the compressor pumping action is unstable, a flow reversal can result. With flow reversal, the volume in the receiver decreases and so does the pressure. The compressor can then, once again, provide positive flow and will pump back into the receiver. This back-and-forth flow motion is called "surging" and can be very damaging to the compressor. In severe cases, surging can cause destruction. While no one can predict exactly the location of the surge point because of manufacturing tolerances and several other factors, manufacturers have a fairly good handle on its location and will generally draw performance curves right to predicted surge. Precaution must be taken to prevent operation at the surge point. This is normally accomplished by recycling flow from the discharge back to the inlet of the compressor. A good rule of thumb is to start the recycle process when the flow through the compressor reaches 110% of the surge flow capacity. This approach essentially yields a 10% safety factor to allow for machining tolerances and instrumentation and valving lag time.

Volume Ratio Effects

What happens when several centrifugal compressor stages are assembled together, each having a characteristic similar to the one just discussed? Will the performance of one stage have any effect on the others?

To answer these questions, refer to Figure 6-6. The performances for three stages are plotted here using percent volume flow and percent head as the abscissa and ordinate of the curves.

Assume a 10% reduction in the mass flow to the compressor. If the inlet conditions are not disturbed, the volume flow to the first stage will drop by 10% also, since the volume flow is directly proportional to the mass flow. What is the percent flow to the second stage? Since the mass flow was reduced by 10%, a 10% reduction in volume flow to the second stage will

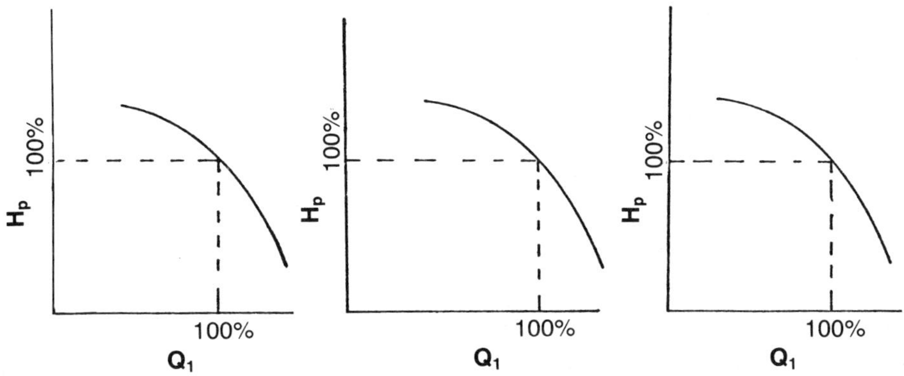

Figure 6-6. Individual impeller characteristics for the three-stage centrifugal compressor.

automatically be seen. However, due to the 10% flow reduction in the first stage, the discharge pressure from the first stage will increase. Refer to the typical performance curve derived in Figure 6-5. Since volume is inversely proportional to pressure, the volume flow to the second stage will be further reduced proportionally to the increase in discharge pressure from the first stage. Therefore, the volume flow to the second stage will be reduced by $(10 + x)\%$ because of a 10% compressor flow reduction. The third stage will have a similar effect; its flow reduction will be $(10 + x + y)\%$.

If a similar analysis is performed to the right of the 100% flow point, a 3% flow increase to the compressor will result in a $(3 + x)\%$ flow increase to the second stage and a $(3 + x + y)\%$ flow increase to the third stage.

Thus, when two or more stages are combined for a compression process, the flow stability of the resultant compressor will be less than the flow stability of any of the individual impellers.

By a similar analysis, one can determine what will happen if a compressor is operated on widely different conditions from those for which it was rated.

Assume a compressor which was rated for a low-molecular-weight gas. If this compressor is used on another process which has identical inlet conditions and volume flow but a significantly higher molecular weight, what happens? Since the volume flow is the same, there will be no effect on the first stage. (For this reason, there are no volume ratio effects for single-stage units, although speeds may have to be adjusted for proper performance.) But what about the second stage? Since the first stage produces the same head in both molecular-weight cases, the discharge pressure from the first stage in the high-molecular-weight case will be higher than it was in the low-molecular-weight case. With a higher pressure, the volume flow entering the second stage will be reduced. Similarly, the higher discharge pressure from the second stage, from considering both the higher molecular weight and the lower flow (i.e. higher head), will further reduce the flow to the third stage.

The converse is not true. Lowering the molecular weight from rated does not increase the flow stability but decreases it also. Consider that lower molecular weights mean lower discharge pressures and therefore higher volume flows. The rated point on one stage may represent stonewall flow on another stage for lower-molecular-weight operation.

The molecular-weight variation in either case can be so great that when one stage is at surge, another is at stonewall, resulting in no flow stability.

It can be deduced from this that when a compressor is operated on a gas with a molecular weight other than that for which the compressor was rated, the flow stability of the compressor will be reduced. The reduction in stability will increase with the variation from rated of the molecular weight.

The effects of wide variations in inlet temperature, inlet compressibility, and *k*-value can be analyzed. These parameters will act to reduce the stability; however, in actual practice the effects of these are not as dramatic as those of the molecular weight. Variations in inlet pressure will have no effect on the flow stability.

One should not get the impression that a compressor can only operate at the point for which it was rated. Compressors do have some degree of flexibility; within reason, effects of stability can be directionally compensated for through speed control. There are limits, however. This should be recognized when attempting to apply a compressor to conditions different from those for which it was rated. If the new conditions are significantly different from the original conditions, a compressor rerate should be considered (refer to Chapter 8).

Curve Parameters

It was discussed in Chapter 4 that the most universal parameter to plot as the independent variable is the inlet, or actual, volume flow. Accepting this, what parameters should normally be plotted against inlet volume flow? There are several (see Figure 6-7):

1. Head (polytropic or adiabatic).
2. Discharge pressure.
3. Power requirement.
4. Efficiency (polytropic or adiabatic).
5. Pressure ratio.
6. Pressure rise.
7. Discharge temperature.

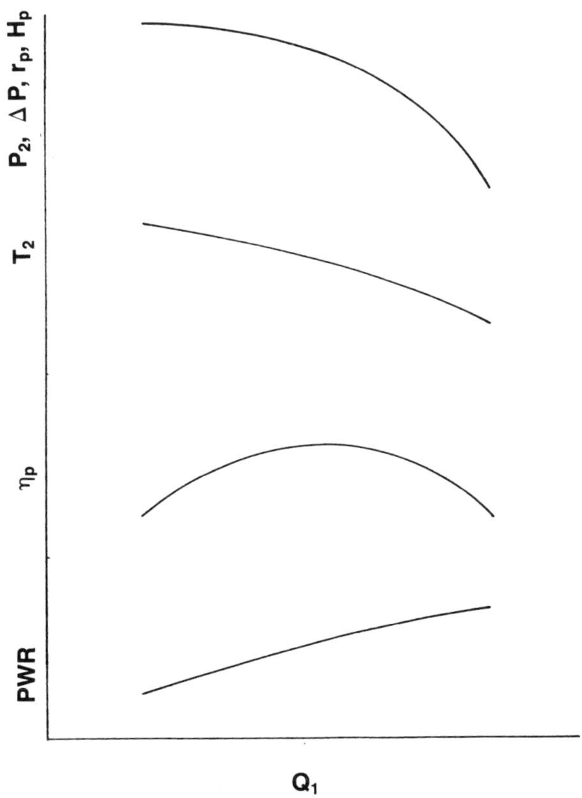

Figure 6-7. Variations of several parameters with inlet volume flow.

Compressor manufacturers can furnish any combination of these performance curves; however, the most common are the first four.

Quadrant curves can also be requested from the manufacturer. These are nomographs. The quadrant curve is used mainly to determine easily the effects on discharge pressure and power requirement caused by variations in the inlet conditions. Figure 6-8 is a typical quadrant curve; there are many other variations. The one shown in Figure 6-8 is useful for variations in volume flow, rotational speed, inlet pressure, inlet temperature, and molecular weight.

Fan Laws

In Chapter 2 it was stated that the head is proportional to the square of the mechanical tip speed:

$$H_p = \frac{\mu u^2}{\text{constant}} \tag{2-17}$$

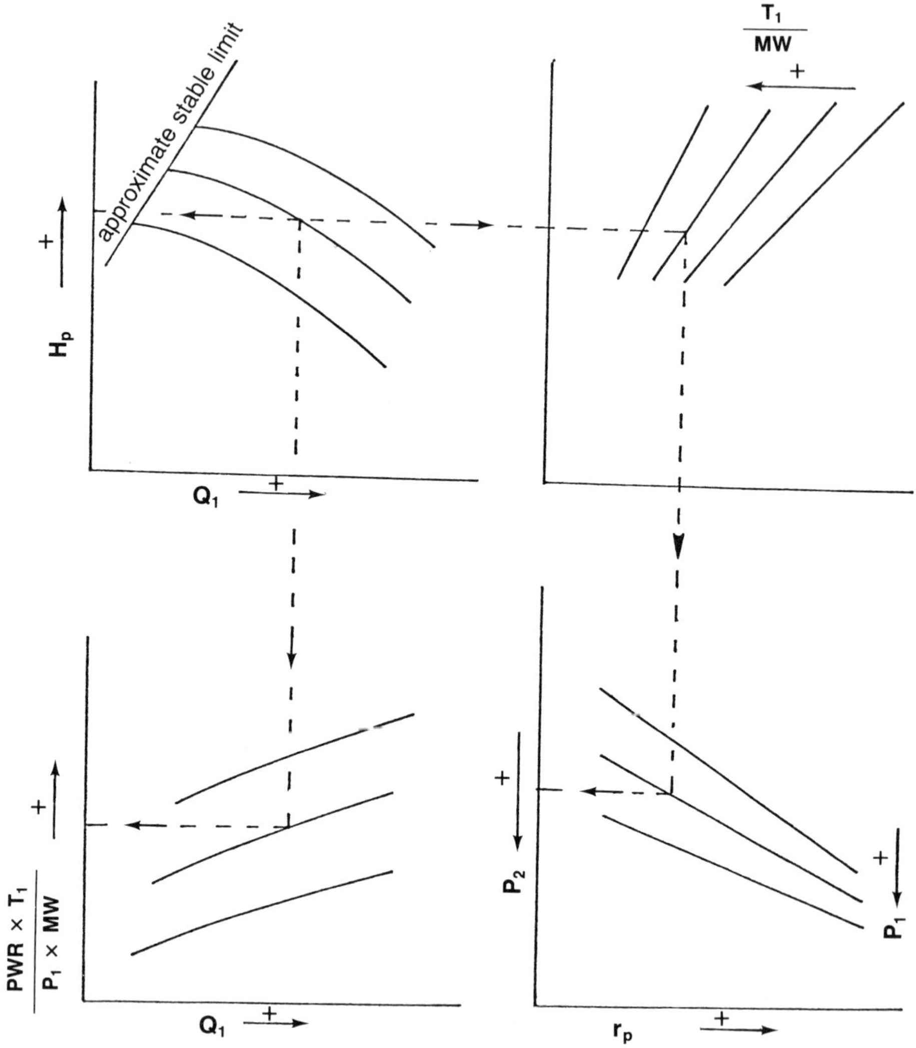

Figure 6-8. Typical quadrant curve for a centrifugal compressor.

The mechanical tip speed is directly proportional to the rotational speed through the following equations:

$$u = \frac{N \pi d}{720} \quad \text{in the English system} \tag{6-1}$$

$$u = \frac{N \pi d}{6 \times 10^4} \quad \text{in the metric system} \tag{6-2}$$

where:

N = rotational speed, RPM
π = pi (3.1416)
d = diameter, in. (English)
= diameter, mm (metric)

Therefore, the head is proportional to the square of the rotational speed:

$$H_p \propto N^2$$

and:

$$\frac{H_{p_n}}{H_{p_o}} = \left(\frac{N_n}{N_o}\right)^2 \tag{6-3}$$

where:

n = new
o = original

The flow through an impeller, and therefore the flow through the compressor, is directly proportional to the rotational speed;

$$Q \propto N$$

and:

$$\frac{Q_n}{Q_o} = \frac{N_n}{N_o} \tag{6-4}$$

If one allows the flow to increase proportionally to the speed and, at the same time, the head to increase proportionally to the speed squared, the resultant power requirement will increase with the cube of the speed:

$$\frac{PWR_n}{PWR_o} = \left(\frac{N_n}{N_o}\right)^3 \tag{6-5}$$

Equations 6-3, 6-4, and 6-5 are known as the fan laws. They are used extensively in compressor performance calculations. Note that the ratios do not allow for volume ratio effects. Therefore, there are limits to the accuracy of the relationships. For single-stage compressors, the fan-law relationships are very accurate. As the number of stages increases, the accuracy of the fan laws deteriorates. One can generally assume that the fan laws will provide very good approximations up to ±10% of the rated speed.

System Resistance

Consider a process whereby the gas entering the compressor is initially at atmospheric pressure. The gas goes through a system of piping, exchangers, and vessels for extraction of a certain product and leaves the system at atmospheric pressure. For this type of process, the compressor is used only to elevate the gas pressure sufficiently to overcome pressure drops due to the piping, exchangers, and vessels. These system pressure drops collectively are known as "system resistance." The system resistance, ΔP, of a system normally varies with the square of the volume flow:

$$\Delta P \propto \Delta Q^2$$

How does a compressor perform in such a system? What does the performance curve look like? Figure 6-9 shows a series of compressor curves for various rotational speeds. Superimposed is a system resistance curve. *The only possible operating points for the compressor are those points where the compressor performance curves intersect the system resistance line.* These are the only points that satisfy both the compressor and the system. Operation at other than these points under steady-state conditions is impossible.

For the purpose of discussion, assume that point A on Figure 6-9 is the rated operating point. If less process flow is desired through the system, how is it obtained? Point B is a lower flow point than point A. To operate at flow point B, a vertical line would be drawn to the system resistance curve. At this point the system resistance is lower than at flow point A. Figure 6-10 shows that process-point B does not fall on compressor speed, N_2, but on a lower compressor speed, N_1.

What has been done? A lower flow has been obtained by decreasing the speed of the compressor. For constant speed drives, the same results could be obtained by throttling the inlet pressure. By a similar analysis, higher flows can be obtained by increasing the speed of the compressor. There are limits to speed increases. The American Petroleum Institute (API) requires

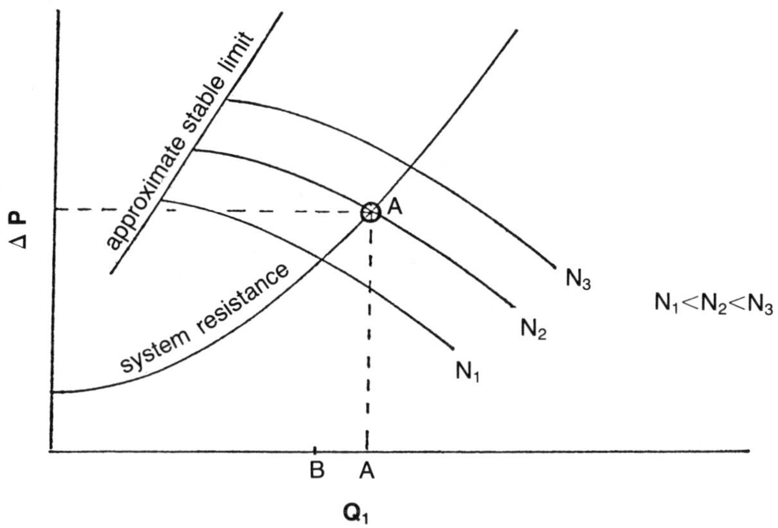

Figure 6-9. Typical system-resistance curve with compressor at rated flow.

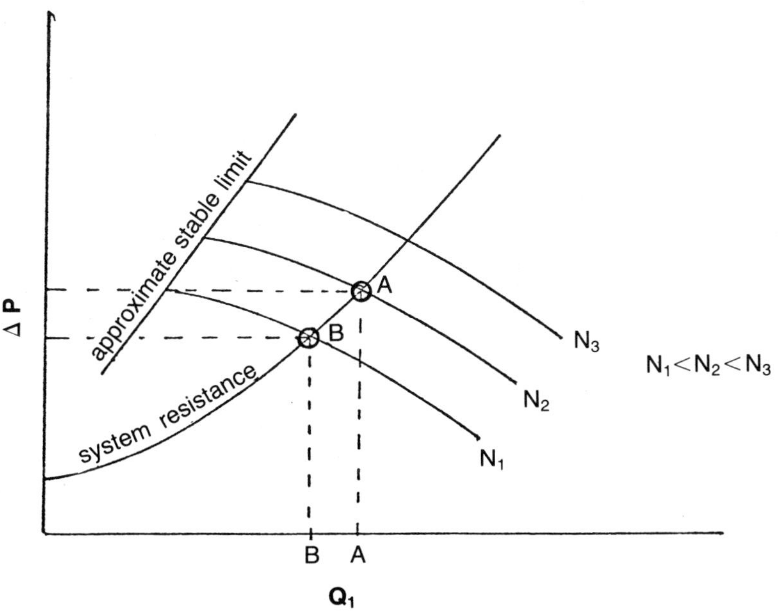

Figure 6-10. Compressor operating at lower than rated flow.

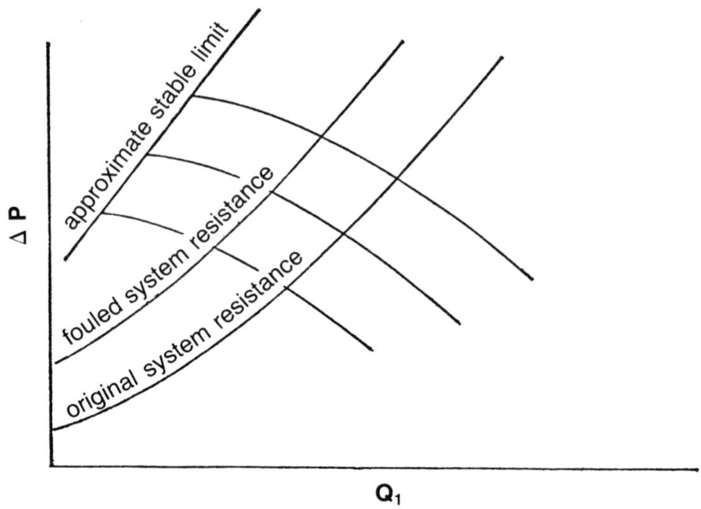

Figure 6-11. Change in system resistance due to system fouling.

that the compressor be capable of 105% of rated speed. This is the value that should be used for speed limitations. The low-end speed limitation is set by process requirements and a consideration of the critical speeds of the unit. A compressor should never be operated within 20% of any critical speed.

Note that with constant-speed drives, flow increases above rated are usually impossible, since there are normally no ways to increase the suction pressure. Therefore, when purchasing constant-speed compressors, precaution must be taken to purchase sufficient polytropic head for the highest possible operating point.

It would appear that it is impossible to surge a properly rated compressor operating in a closed system governed by system resistance. However, as the process operates over a period of time, it may begin to foul the system, thereby increasing the system pressure drop, as shown in Figure 6-11. If sufficient fouling occurs, it is entirely possible to drive the compressor into surge unless protected by recycle. Note that with increased fouling, it also becomes increasingly more difficult to deliver rated capacity.

Now that the compressor performance curve has been reviewed at some length, the next chapter will be devoted to methods of obtaining the most use from a manufacturer-supplied performance curve.

7

How to Use Performance Curves*

What happens when the compressor inlet conditions vary from those for which the compressor was rated? Is the manufacturer's curve still valid? If so, how is it used or modified?

It was shown in Chapter 2 that there are many advantages to using polytropic head over adiabatic head. However, it was also stated that because single-stage compressors only have one stage and because the pressure ratios are low, adiabatic head works well for single-stage applications.

Chapter 6 discussed volume ratio effects and how a significant variation in inlet conditions can seriously reduce the stability of multi-stage compressors.

This chapter will develop procedures to apply to performance curves for varying inlet conditions. Note that volume ratio effects cannot be considered without the manufacturer's stage data. Therefore, the approach here will be to ignore volume ratio effects and develop procedures for single-stage compressors, with the understanding that these can be applied to multi-stage compressors with caution. Furthermore, the procedures will be developed using adiabatic head, since single-stage performance curves are usually presented this way. The procedures thus developed will be directly applicable for polytropic head.

Note that a variation in inlet conditions does not change the head output of an impeller. The head output is a function of the impeller tip speed only,

---

*Adapted by special permission from *Chemical Engineering*, Jan. 25, 1982, ©1982 by McGraw-Hill, Inc., New York, NY 10020; and *CAGI Technical Digest*, vol. 9, no. 2, by Compressed Air and Gas Institute, Cleveland, Ohio 44115.

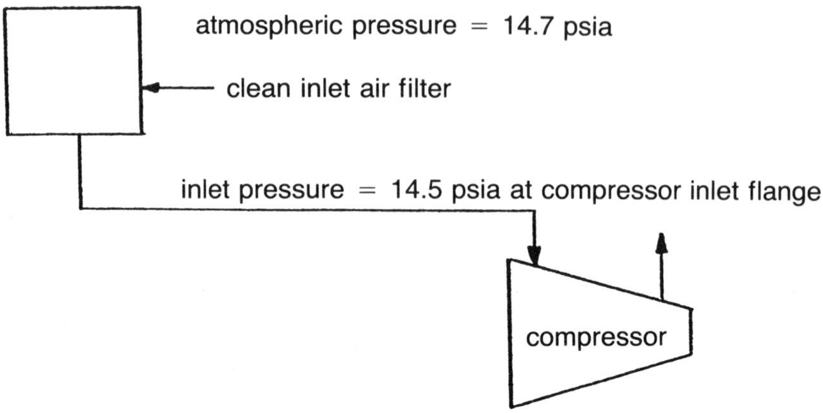

Figure 7-1. Model of air compressor with clean inlet air filter.

ignoring volume ratio effects. However, the inlet conditions will affect the discharge pressure. Therefore, most of this discussion will be conducted using a volume flow versus discharge pressure curve. The procedures will be developed in the English system; however, the results can be converted to the metric system.

For single-stage compressors, the inlet compressibility should be used to calculate the head; for multi-stage compressors, the average compressibility should be used. In this chapter, Z will be used for compressibility, with the understanding that it refers to inlet for single-stage compressors, and to average for multi-stage compressors.

Variation in Inlet Pressure

First consider a variation in inlet pressure. For the purpose of discussion, consider that the compressor draws atmospheric air through an inlet filter. The model of such a system is shown in Figure 7-1. The manufacturer furnished two performance curves for the compressor: Figure 7-2, which shows the adiabatic head, and Figure 7-3, which shows the discharge pressure. The rated compressor inlet pressure, as tabulated on the performance curve, is 14.5 psia. When the filter becomes dirty (Figure 7-4), the inlet pressure at the compressor flange drops to 14.2 psia. What is the effect on discharge pressure and power requirement at the compressor rated inlet volume flow?

The discharge pressure is related to the adiabatic head by:

$$H_{ad} = ZRT_1 \frac{k}{k-1} [r_p^{(k-1)/k} - 1] \tag{7-1}$$

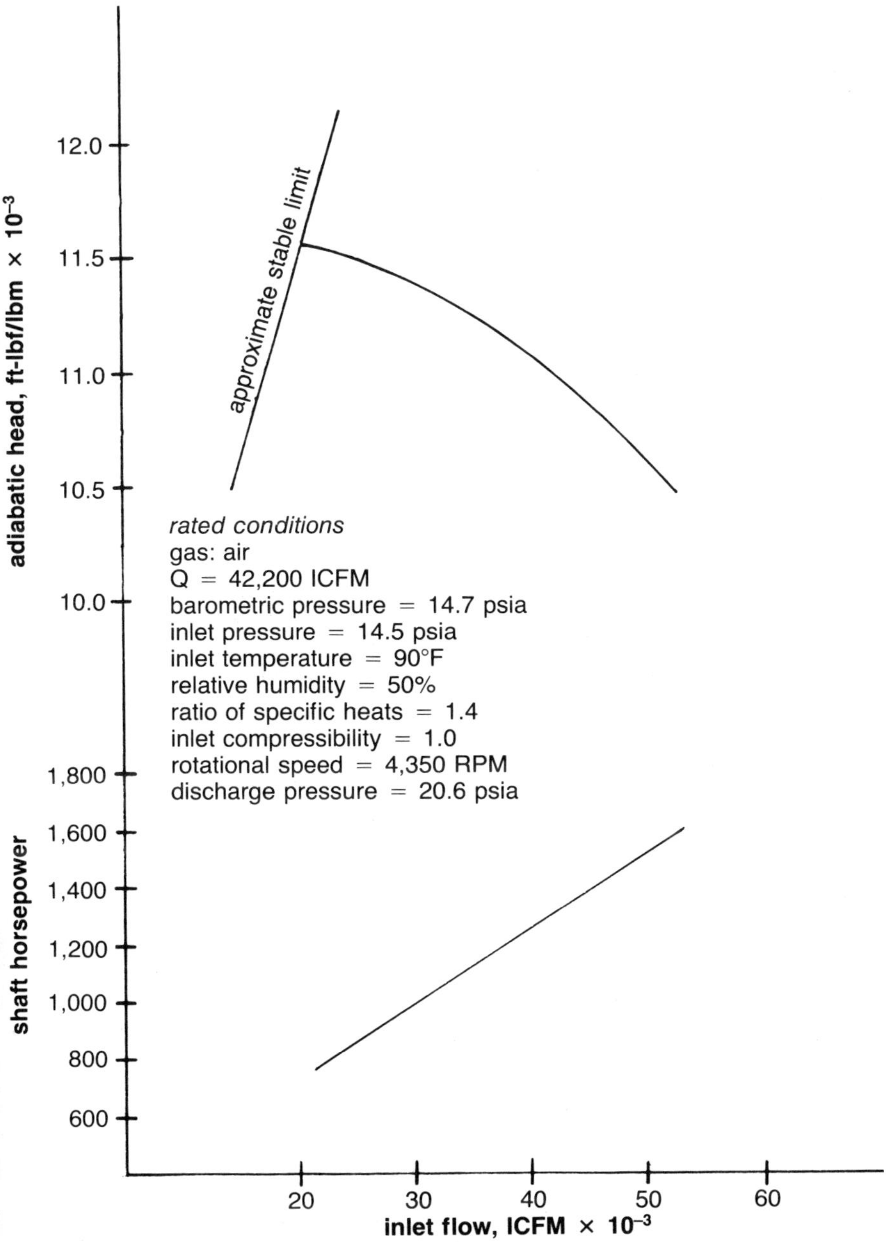

Figure 7-2. Typical single-stage centrifugal compressor performance curve for adiabatic head and shaft power versus inlet volume flow.

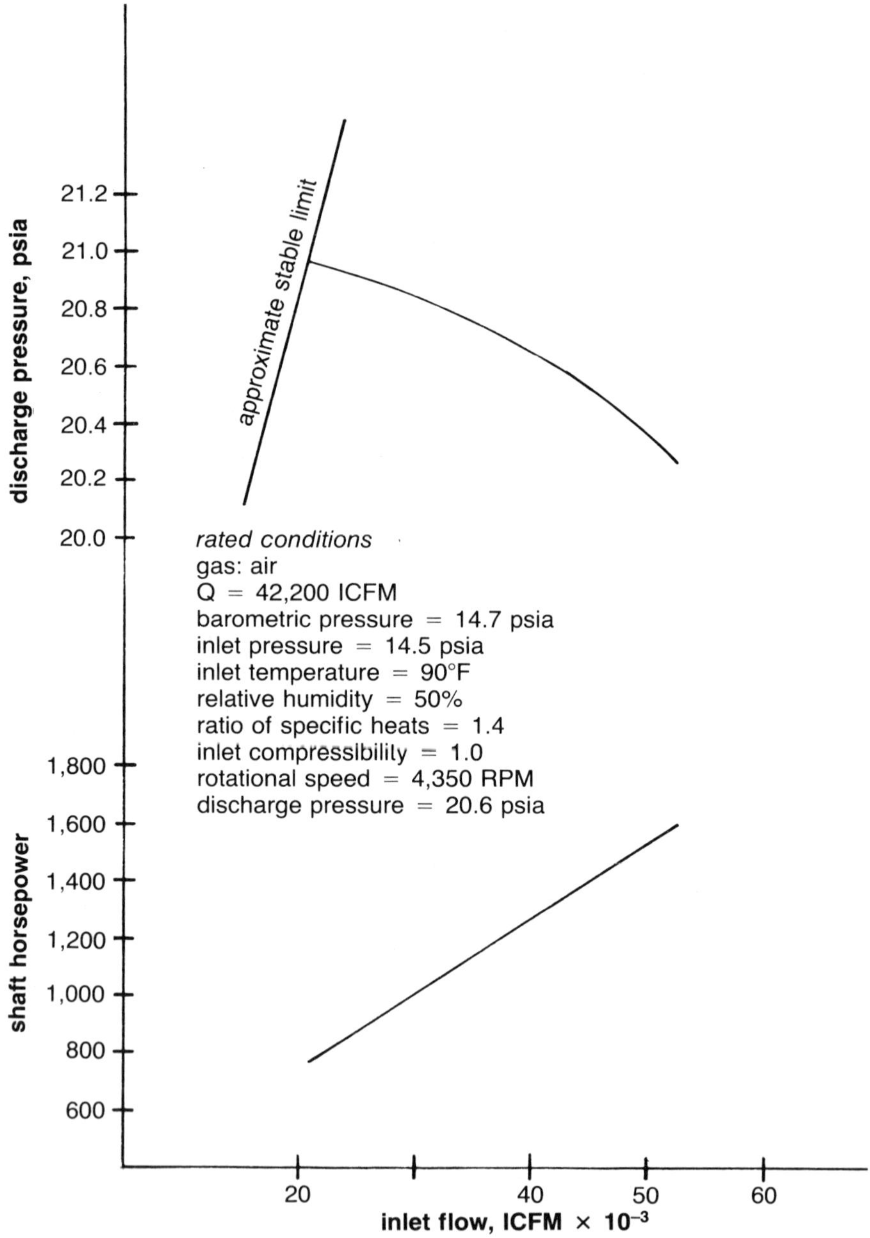

Figure 7-3. Typical single-stage centrifugal compressor performance curve for discharge pressure and shaft power versus inlet volume flow.

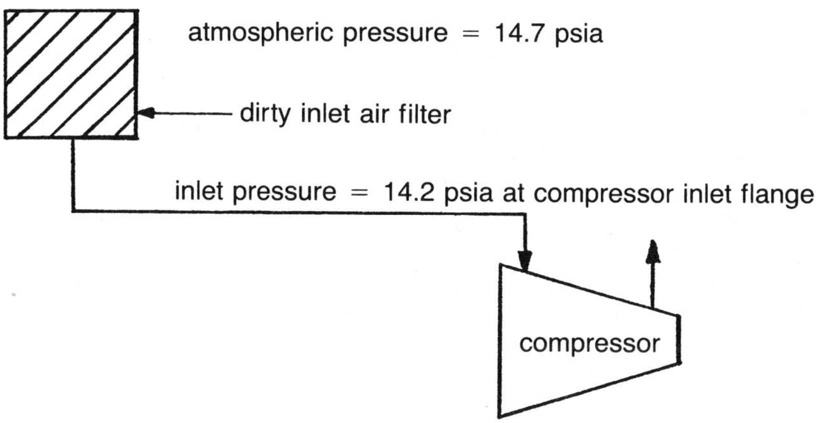

Figure 7-4. Model of air compressor with dirty inlet air filter.

For a given volume flow at a given rotational speed, the head output is constant and since there is no variation in the other inlet conditions, the pressure ratio does not vary. Therefore:

$$r_p = (r_p)_r = \frac{20.6}{14.5} = 1.42$$

$$P_2 = P_1(r_p)_r \tag{7-2}$$

$$= 14.2\,(1.42) = 20.2 \text{ psia}$$

where the subscript r stands for rated. The power requirement is given by:

$$PWR = \frac{mH_{ad}}{33{,}000\ \eta_{ad}} + L_m \tag{7-3}$$

The volume flow is related to the mass flow by:

$$m = \frac{Q_1}{v_1} = \frac{144\ Q_1 P_1}{ZRT_1} \tag{7-4}$$

By comparing Equations 7-3 and 7-4, it can be seen that:

$$PWR \propto m \propto P_1$$

and:

$$PWR = \left[\frac{P_1}{(P_1)_r}\right] PWR_r \tag{7-5}$$

Note that this relationship is not entirely true, since the power requirement is composed of gas power and mechanical losses. The mechanical losses are constant for a given speed, but are generally a small portion of the total power requirement. Therefore, this procedure will usually yield a useful approximation.

The performance curve indicates that the power requirement for the rated conditions is 1,315 hp. Therefore:

$$PWR = \left(\frac{14.2}{14.5}\right)(1{,}315) = 1{,}290 \text{ hp}$$

So far, system resistance as discussed in Chapter 6 has been ignored. For many applications, system resistance effects are small and the foregoing analysis holds. What about the applications where system resistance governs the operation of the compressor? Figure 7-5 shows a typical system resistance line superimposed on two compressor performance curves. The solid curve is for compressor rated inlet conditions; the lower curve shows the effects of a reduction in inlet pressure only. Point A on the curve is the rated compressor operating point.

The foregoing assumed constant inlet volume flow; therefore, the discharge pressure was calculated at point C. The power requirement calculated was also for operation at point C. If this compressor operation was governed by system resistance, the compressor would search its new performance curve (the lower one in Figure 7-5) until the performance intersected system requirements. The result would be operation at point B. Therefore, the inlet volume flow would be somewhat less than rated and the discharge pressure slightly higher than calculated. Reference to the shape of the power curve on Figures 7-2 and 7-3 shows that the power requirement would be less than calculated.

Variation in Inlet Temperature

Suppose the inlet temperature drops to 40°F and all other inlet conditions remain at rated values. What is the effect on discharge pressure and power requirement of the compressor defined by Figure 7-2 at the rated inlet volume flow?

Rearranging Equation 7-1 yields:

$$r_p^{(k-1)/k} - 1 = \frac{H_{ad}(k-1)}{ZRT_1 k}$$

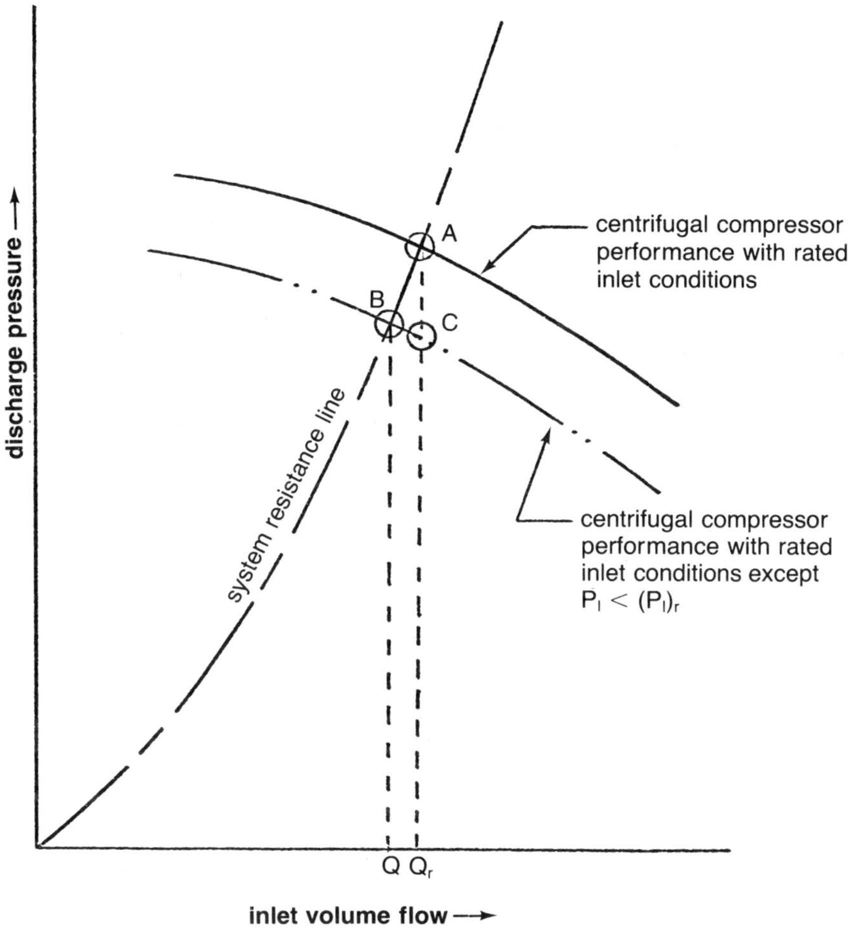

Figure 7-5. Effect on centrifugal compressor performance due to a change in inlet pressure.

which implies that a change in inlet temperature inversely affects the pressure ratio. Therefore, for a change in inlet temperature only:

$$(r_p)_r^{(k-1)/k} - 1 = \frac{C_1}{(T_1)_r}$$

where:

$$C_1 = \frac{H_{ad}(k - 1)}{ZRk} = \text{constant}$$

$$r_p^{(k-1)/k} - 1 = \frac{C_1}{T_1}$$

$$\left[\frac{r_p^{(k-1)/k} - 1}{(r_p)_r^{(k-1)/k} - 1}\right] = \frac{(T_1)_r}{T_1}$$

$$r_p = \left[\frac{(T_1)_r}{T_1}\{(r_p)_r^{(k-1)/k} - 1\} + 1\right]^{k/(k-1)}$$

$$P_2 = P_1\left[\frac{(T_1)_r}{T_1}\{(r_p)_r^{(k-1)/k} - 1\} + 1\right]^{k/(k-1)} \tag{7-6}$$

$$= 14.5\left[\frac{550}{500}(1.42^{.286} - 1) + 1\right]^{3.5} = 21.3 \text{ psia}$$

By comparing Equations 7-3 and 7-4, it can be seen that:

$$PWR \propto \frac{1}{T_1}$$

$$PWR = \left(\frac{(T_1)_r}{T_1}\right) PWR_r \tag{7-7}$$

$$= \left(\frac{550}{500}\right)(1{,}315) = 1{,}450 \text{ hp}$$

Here, again, the effects of system resistance have been neglected. For those applications where system resistance governs, refer to Figure 7-6. The solid curve represents compressor performance for rated inlet conditions; the higher curve represents a drop in inlet temperature, with all other inlet conditions at rated values. Point A is the compressor rated point. A constant inlet volume flow was considered; therefore, the performance was calculated at point C. The compressor will search its new performance curve until it reaches point B, the point of intersection with the system resistance line. As shown on Figure 7-6, the inlet flow will be somewhat higher than rated and the discharge pressure slightly lower than calculated. Reference to Figures 7-2 or 7-3 will show that the power requirement will be higher than calculated.

Comparing Figures 7-5 and 7-6 will show that a drop in inlet pressure will lower the discharge pressure curve, while a drop in inlet temperature will raise the curve. From this, it can be deduced that rated performance

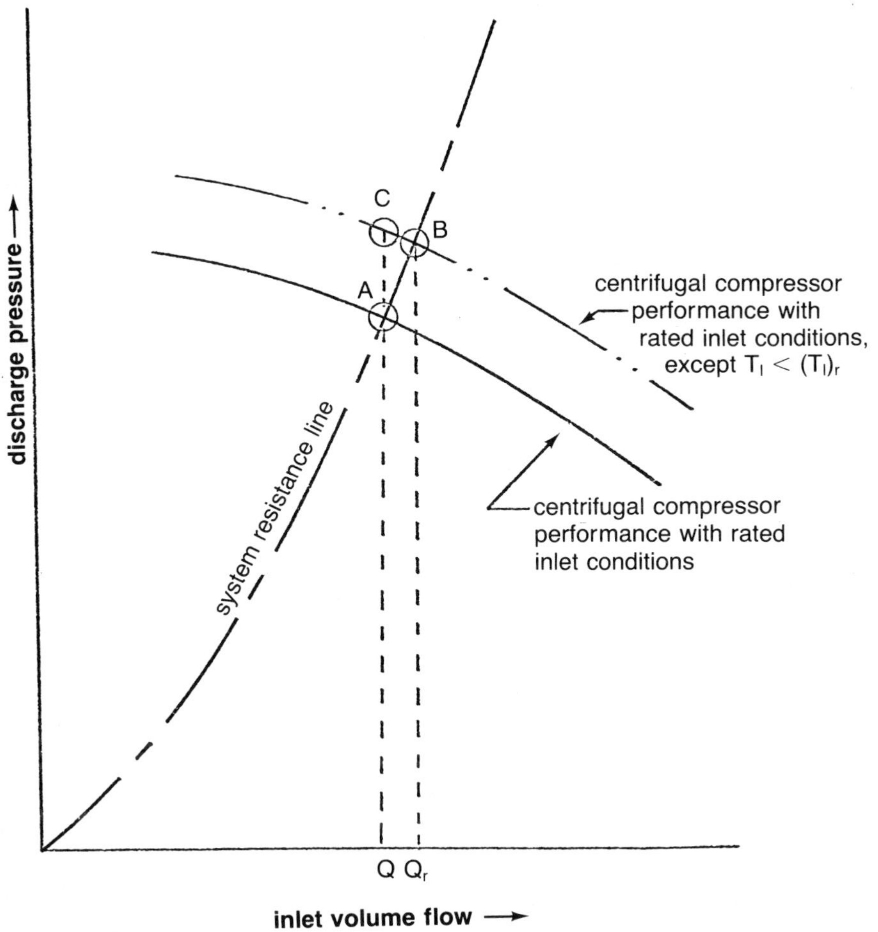

Figure 7-6. Effect on centrifugal compressor performance due to a change in inlet temperature.

can be obtained on cold days by suction throttling the inlet pressure. (Note: It is also possible to obtain the same result by lowering the speed on variable-speed drives. This will be covered later in this chapter.) Suction throttling also lowers the power requirement, since the power is directly proportional to the inlet pressure.

Since, for this discussion, the approach is to analyze each variable independently, the change in water content in the air due to the change in inlet temperature has not been taken into account. For the rated conditions, the molecular weight is 28.7. When the inlet temperature reaches 40°F, the molecular weight becomes 28.9, as long as the relative humidity remains at 50%.

The effects of variation in water content will be considered next.

Variation in Molecular Weight

The molecular weight of an air-water vapor mixture varies with composition. The effect of this variation is usually small for air-water vapor mixtures and can be ignored. However, gas compressors can operate over a wide range of molecular weights, making this variable quite significant. To illustrate how variation in molecular weight might be accounted for, consider a change in relative humidity at the rated inlet temperature, and therefore a change in air-water vapor composition.

Consider that on a given day the atmospheric temperature is 90°F and the relative humidity 100%. For a barometric pressure of 14.7 psia, the molecular weight is approximately 28.4. It was stated in the previous section that the molecular weight at the rated conditions is 28.7.

First investigate how the molecular weight enters the head equation. The term R in Equation 7-1 is given by:

$$R = \frac{1{,}545}{MW}$$

Therefore, one can rewrite Equation 7-1 as:

$$H_{ad} = Z\left(\frac{1{,}545}{MW}\right) T_1 \left(\frac{k}{k-1}\right)[r_p^{(k-1)/k} - 1] \tag{7-8}$$

and rearrange as:

$$r_p^{(k-1)/k} - 1 = \frac{H_{ad}\, MW\, (k-1)}{1{,}545\, Z\, T_1\, k}$$

which implies that a change in molecular weight directly affects the pressure ratio. Therefore, for a change in molecular weight only:

$$(r_p)_r^{(k-1)/k} - 1 = C_2\, MW_r$$

where:

$$C_2 = \frac{H_{ad}\,(k-1)}{1{,}545\, Z\, T_1\, k} = \text{constant}$$

$$r_p^{(k-1)/k} - 1 = C_2\, MW$$

$$r_p = \left[\frac{MW}{MW_r}\{(r_p)_r^{(k-1)/k} - 1\} + 1\right]^{k/(k-1)}$$

$$P_2 = P_1\left[\frac{MW}{MW_r}\{(r_p)_r^{(k-1)/k} - 1\} + 1\right]^{k/(k-1)} \tag{7-9}$$

$$= 14.5\left[\left(\frac{28.4}{28.7}\right)(1.42^{.286} - 1) + 1\right]^{3.5} = 20.5 \text{ psia}$$

By comparing Equations 7-3, 7-4, and 7-8, it can be seen that:

$$PWR \propto MW$$

$$PWR = \frac{MW}{MW_r}(PWR_r) \tag{7-10}$$

$$= \left(\frac{28.4}{28.7}\right)(1{,}315) = 1{,}300 \text{ hp}$$

Figure 7-7 illustrates the effects of system resistance on this analysis. Here, it can be seen that a reduction in molecular weight lowers the compressor performance curve. Thus, the result of a molecular weight variation is directionally the same as a variation in inlet pressure.

Variation in Inlet Compressibility and Ratio of Specific Heats

For an air process, the variation in these two parameters is so slight that they can be ignored. For gases other than air, however, the variation in these parameters may prove significant.

Figure 7-8 shows the directional effect of a variation in both Z only and k only. The dotted curve indicates that a decrease in inlet compressibility raises the curve (when inlet volume flow is plotted against discharge pressure); the dashed curve indicates that a decrease in the ratio of specific heats also raises the performance curve.

Once again, one should realize that this entire analysis has been based on the fact that for a given rotational speed, each inlet volume flow is associated with one and only one head value. *The head curve does not change for any of the preceding discussions.*

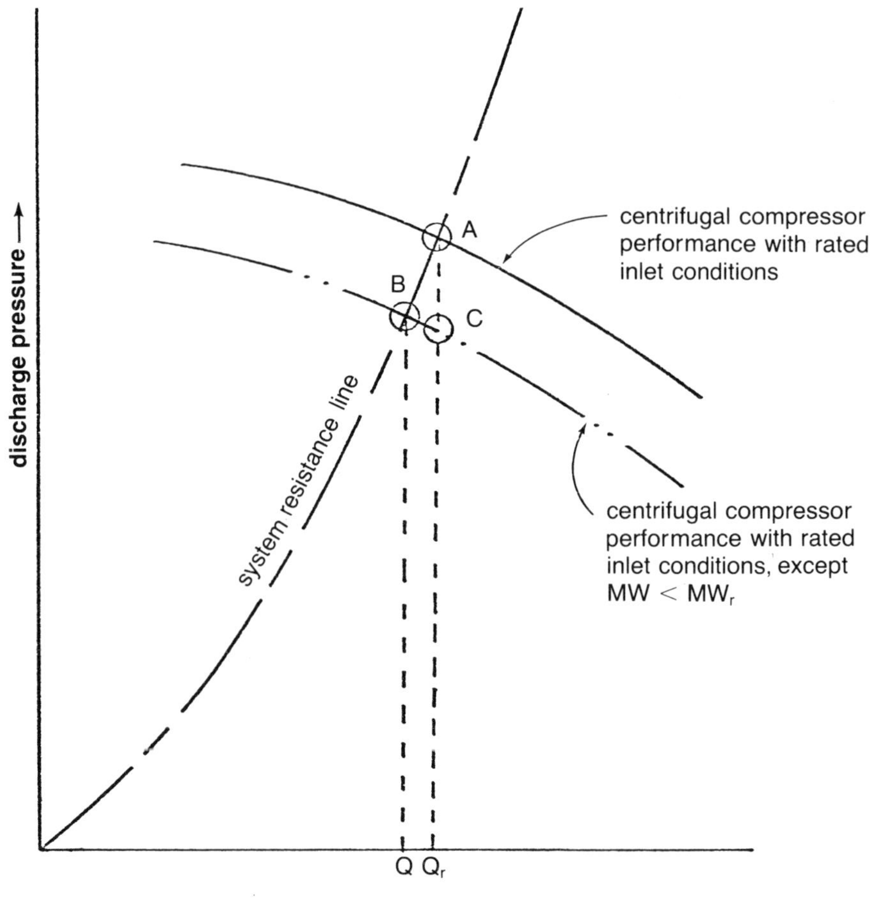

Figure 7-7. Effect on centrifugal compressor performance due to a change in molecular weight.

A General Approach for Constant Rotational Speed

A general equation can be formulated to determine the discharge pressure which results from a variation in any or all of the inlet conditions at rated inlet volume flow for constant rotational speed. Referring to the derivation of Equation 7-6 for a variation in inlet temperature only, it can be seen that a similar approach would yield a general discharge pressure equation for inlet condition variations.

First assume that the ratio of specific heats and inlet volume flow are constant; the discharge pressure equation then becomes:

$$P_2 = P_1 \left\{ \left[\frac{(T_1)_r}{T_1} \right] \left[\frac{Z_r}{Z} \right] \left[\frac{MW}{MW_r} \right] [(r_p)_r^{(k-1)/k} - 1] + 1 \right\}^{k/(k-1)} \tag{7-11}$$

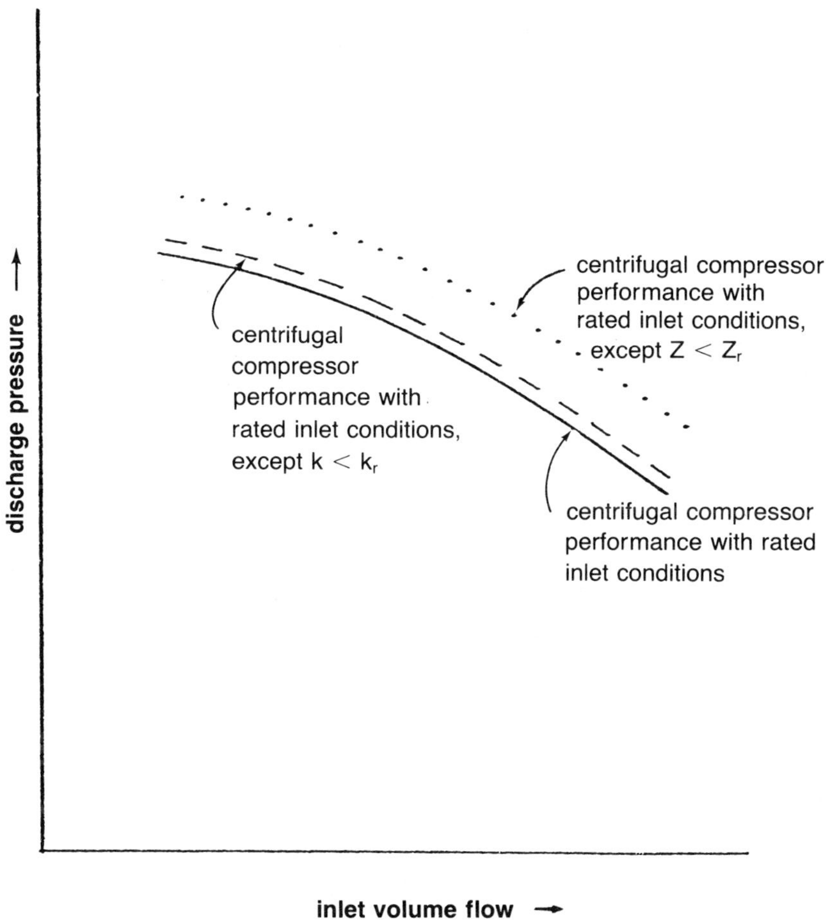

Figure 7-8. Effects on centrifugal compressor performance due to a change in inlet compressibility only, and a change in ratio of specific heats only.

If the ratio of specific heats is allowed to vary:

$$P_2 = P_1 \left\{ \left[\frac{(T_1)_r}{T_1}\right]\left[\frac{Z_r}{Z}\right]\left[\frac{MW}{MW_r}\right]\left[\frac{k}{k-1}\right]_r \left[\frac{k-1}{k}\right] \cdot [(r_p)_r^{[(k-1)/k]_r} - 1] + 1 \right\}^{k/(k-1)} \tag{7-12}$$

In either case, the equation for power requirement is:

$$PWR = \left[\frac{P_1}{(P_1)_r}\right]\left[\frac{(T_1)_r}{T_1}\right]\left[\frac{Z_r}{Z}\right]\left[\frac{MW}{MW_r}\right] PWR_r \tag{7-13}$$

Note that since k does not vary much for air-water vapor mixtures, it can be considered constant for these applications, with little or no error introduced. The variation in k can be quite significant for gas compressors, however; the reader will have to consider its effect for certain applications.

Consider the single-stage compressor defined by Figure 7-2. For discussion, determine the discharge pressure and power requirement at the rated inlet volume flow for a variation of inlet conditions:

$$P_1 = 14.2 \text{ psia}$$

$$T_1 = 40°\text{F} = 500°\text{R}$$

$$MW = 28.4$$

Since the ratio of specific heats is constant, use Equation 7-11:

$$P_2 = (14.2)\left[\left(\frac{550}{500}\right)\left(\frac{1.0}{1.0}\right)\left(\frac{28.4}{28.7}\right)(1.42^{.286} - 1) + 1\right]^{3.5} = 20.8 \text{ psia}$$

$$PWR = \left(\frac{14.2}{14.5}\right)\left(\frac{550}{500}\right)\left(\frac{1.0}{1.0}\right)\left(\frac{28.4}{28.7}\right)(1{,}315) = 1{,}400 \text{ hp}$$

When necessary, use the system resistance curve to make the proper adjustments.

The Constant Weight Flow Process

The preceding discussions were limited to constant inlet volume flow. However, sometimes it is necessary to consider constant weight flow. The performance curve can be used to predict discharge pressure and power requirement for this process also.

For a constant weight flow process, a variation in inlet conditions will affect the inlet volume flow. Refer to Equation 7-4. Since in this case, inlet volume flow varies, the head produced by the impeller will vary also.

To illustrate the procedure, use the compressor defined by Figure 7-2 and predict the discharge pressure for a change in both inlet temperature and inlet pressure. The procedure thus developed is applicable to Figure 7-3.

Consider the following: inlet temperature at 100°F, inlet pressure at 14.0 psia, and molecular weight, ratio of specific heats, and inlet compressibility at rated values.

Equation 7-4 implies that for a constant weight flow process:

$$Q_1 \propto \frac{T_1}{P_1}$$

$$Q_1 = \left(\frac{T_1}{(T_1)_r}\right)\left(\frac{(P_1)_r}{P_1}\right)(Q_1)_r \tag{7-14}$$

$$= \left(\frac{560}{550}\right)\left(\frac{14.5}{14.0}\right)(42{,}200) = 44{,}500 \text{ ICFM}$$

Reference to Figure 7-2 will show that at the rated inlet volume flow of 42,200 ICFM, the head produced is 11,000 ft-lbf/lbm. The head produced at the new flow of 44,500 ICFM is 10,900 ft-lbf/lbm. It can be deduced that:

$$(r_p^{(k-1)/k} - 1) \propto \frac{H_{ad}}{T_1}$$

$$P_2 = P_1 \left\{\left[\frac{H_{ad}}{(H_{ad})_r}\right]\left[\frac{(T_1)_r}{T_1}\right][(r_p)_r^{(k-1)/k} - 1] + 1\right\}^{k/(k-1)} \tag{7-15}$$

$$= 14.0\left[\left(\frac{10{,}900}{11{,}000}\right)\left(\frac{550}{560}\right)(1.42^{.286} - 1) + 1\right]^{3.5}$$

$$= 19.7 \text{ psia}$$

The power requirement for the rated inlet conditions at a flow of 44,500 ICFM can be determined from the performance curve. This value is 1,380 hp. From Equation 7-13:

$$PWR = \left(\frac{14.0}{14.5}\right)\left(\frac{550}{560}\right)(1{,}380) = 1{,}310 \text{ hp}$$

The procedure for using Figure 7-3 follows from this. In this case, the discharge pressure at the new inlet volume flow for the rated inlet conditions is derived directly. This new discharge pressure along with the rated inlet pressure defines the new ''rated'' pressure ratio. Apply the temperature correction to the pressure ratio and calculate the discharge pressure as before. For example, the discharge pressure from Figure 7-3 at 44,500 ICFM and 90°F inlet temperature is 20.54 psia. Consider this new discharge pressure "rated" and use the temperature correction with the new inlet pressure to determine the final discharge pressure:

$$P_2 = (14.0)\left\{\left(\frac{550}{560}\right)\left[\left(\frac{20.54}{14.5}\right)^{.286} - 1\right] + 1\right\}^{3.5} = 19.7 \text{ psia}$$

Variation in Rotational Speed

Centrifugal compressor performance varies with the rotational speed. If the speed variation is not too large—say, 90%–105% rated speed—compressor performance can be predicted from the fan laws.

The fan-law relationships are:

$$H_{ad} = (H_{ad})_r \left(\frac{N}{N_r}\right)^2 \tag{7-16}$$

$$Q_1 = (Q_1)_r \left(\frac{N}{N_r}\right) \tag{7-17}$$

From these relationships, one can predict the speed required to overcome the effects of inlet condition variations. The following procedure can be used for any inlet condition variation and for any type of performance curve.

Consider the previously presented case of variation in inlet pressure only. The drop in inlet pressure resulted in a lower discharge pressure of 20.2 psia. What speed is required to raise the discharge pressure to the rated value of 20.6 psia at rated inlet volume flow?

First determine the pressure ratio at rated speed for the inlet conditions under consideration by using the applicable procedures from the foregoing discussions. For this case, the pressure ratio is:

$$(r_p)_{ic} = \frac{20.2}{14.2} = 1.42$$

Then determine the required pressure ratio:

$$(r_p)_{\text{req}} = \frac{20.6}{14.2} = 1.45$$

For any given set of inlet conditions:

$$r_p^{(k-1)/k} - 1 \propto H_{ad}$$

$$H_{ad} \propto N^2$$

$$r_p^{(k-1)/k} - 1 \propto N^2$$

Therefore, the following rotational speed equation can be derived:

$$N_{\text{req}} = N_r \left[\frac{(r_p^{(k-1)/k} - 1)_{\text{req}}}{(r_p^{(k-1)/k} - 1)_{ic}} \right]^{1/2} \tag{7-18}$$

$$= (4{,}350) \left[\frac{1.45^{.286} - 1}{1.42^{.286} - 1} \right]^{1/2} = 4{,}490 \text{ RPM}$$

At 4,490 RPM, the compressor will produce the required head; however, the speed increase will affect the volume flow also. The new flow will be:

$$Q = 42{,}200 \left(\frac{4{,}490}{4{,}350} \right) = 43{,}560 \text{ ICFM}$$

Reduction of flow back to rated at the new speed of 4,490 RPM would result in too much head.

Figure 7-9 shows the effect on inlet volume flow and head due to an increase in rotational speed. The solid curve represents rated conditions; point A is the rated operating point. The dashed curve represents compressor performance at the previously calculated speed of 4,490 RPM. Point B represents the desired head; however, the inlet volume flow has also been increased. Reduction of inlet volume flow to rated will result in performance at point C and, therefore, too much head.

The volume flow effect can best be handled by determining the percentage change in head that results from a change in relative loading.

$$\%\, Q = \frac{(Q_1)_{\text{req}} N_r}{(Q_1)_r N_{\text{req}}} \tag{7-19}$$

$$= \frac{(42{,}200)(4{,}350)}{(42{,}200)(4{,}490)} = 0.969 = 96.9\%$$

Refer to Figure 7-2 to determine the percentage change in head that results in going from 100% $(Q_1)_r$ to 96.9% $(Q_1)_r$:

100% $(Q_1)_r = 42{,}200$

96.9% $(Q_1)_r = 40{,}900$

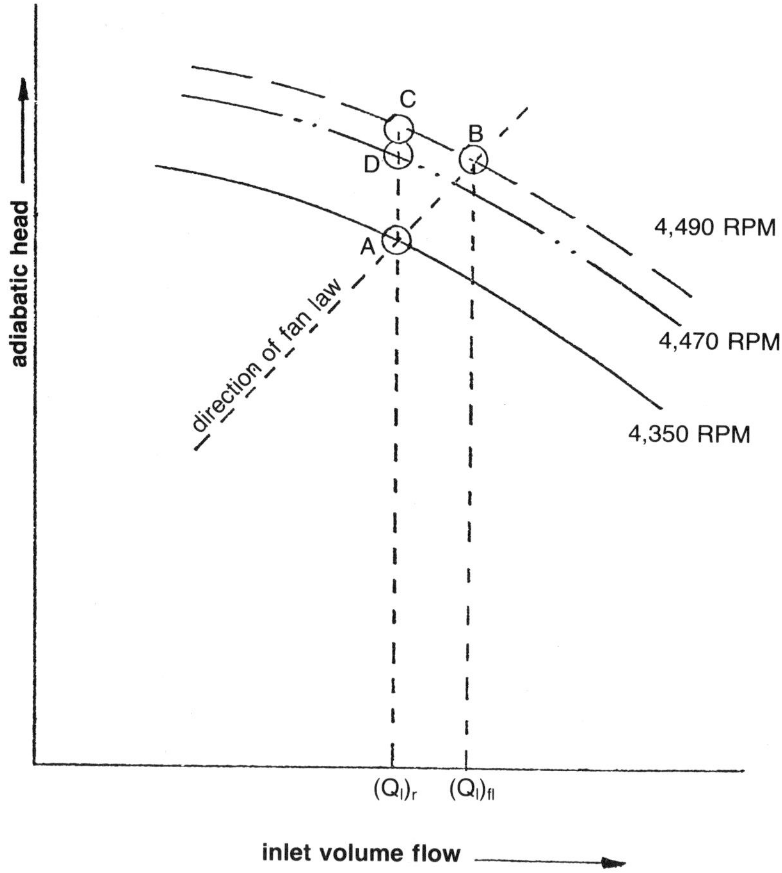

Figure 7-9. Effect on centrifugal compressor performance due to a change in rotational speed.

At 42,200 ICFM, H_{ad} = 11,000 ft-lbf/lbm. At 40,900 ICFM, H_{ad} = 11,075 ft-lbf/lbm. Then:

$$\% \, H_{ad} = \frac{11{,}075}{11{,}000} = 1.007 = 100.7\%$$

Equation 7-18 can now be revised as follows:

$$N_{\text{req}} = N_r \left[\frac{(r_p^{(k-1)/k} - 1)_{\text{req}}}{(r_p^{(k-1)/k} - 1)_{ic} \, (\% \, H_{ad})} \right]^{1/2} \qquad (7\text{-}20)$$

$$= 4{,}350 \left[\frac{1.45^{.286} - 1}{(1.42^{.286} - 1)(1.007)} \right]^{1/2} = 4{,}470 \text{ RPM}$$

The result is operation at point D, the desired operating point.

The foregoing discussion used a constant inlet volume flow of 42,200 ICFM. The procedure is valid for variable inlet volume flow. For example, if the inlet volume flow was to be reduced to 40,000 ICFM, the $\% \, Q$ would become:

$$\% \, Q = \left(\frac{40{,}000}{42{,}200}\right)\left(\frac{4{,}350}{4{,}490}\right) = 0.918 = 91.8\%$$

The rest of the procedure is the same.

The last step in analyzing speed variation may seem difficult with a performance curve with discharge pressure. However, when using a curve such as Figure 7-3, remember that:

$$\% \, H_{ad} = \% \, (r_p^{(k-1)/k} - 1)$$

for any given set of inlet conditions.

For this case, it would be easier to estimate the power requirement by calculating the adiabatic heads and mass flows for the rated conditions and the required conditions. Then, using Equation 7-21:

$$PWR = \left[\frac{(H_{ad})_{\text{req}}}{(H_{ad})_r}\right]\left[\frac{m_{\text{req}}}{m_r}\right] PWR_r \qquad (7\text{-}21)$$

Equation 7-21 will provide a fairly good estimate if $\% \, Q$ ranges from 95% to 105%. An even better approximation can be obtained if a variable speed curve is received from the manufacturer, since then the required speed line can be drawn and the inlet conditions corrected for by Equation 7-13. (Note that the manufacturer's variable speed curve will incorporate any speed effects on the flow stability, slope of the characteristic curve, and location of the optimum efficiency point.)

Summary

The preceding techniques were derived from the adiabatic head equation; they apply equally well to the polytropic head equation. For utility, some of the equations previously derived follow.

For constant rotational speed, constant volume flow, and constant k-value:

$$P_2 = P_1 \left\{\left[\frac{(T_1)_r}{T_1}\right]\left[\frac{Z_r}{Z}\right]\left[\frac{MW}{MW_r}\right][(r_p)_r^{(k-1)/k} - 1] + 1\right\}^{k/(k-1)} \qquad (7\text{-}11)$$

For constant rotational speed, constant volume flow, and variable k-value:

$$P_2 = P_1 \left\{ \left[\frac{(T_1)_r}{T_1}\right] \left[\frac{Z_r}{Z}\right] \left[\frac{MW}{MW_r}\right] \left[\frac{k}{k-1}\right]_r \left[\frac{k-1}{k}\right] \cdot [(r_p)_r^{[(k-1)/k]_r} - 1] + 1 \right\}^{k/(k-1)} \tag{7-12}$$

The equation for the power requirement for constant inlet volume flow and constant speed is:

$$PWR = \left[\frac{P_1}{(P_1)_r}\right] \left[\frac{(T_1)_r}{T_1}\right] \left[\frac{Z_r}{Z}\right] \left[\frac{MW}{MW_r}\right] PWR_r \tag{7-13}$$

For any given set of inlet conditions, the speed required to provide a given pressure ratio at a given inlet volume flow is as follows:

$$N_{req} = N_r \left[\frac{(r_p^{(k-1)/k} - 1)_{req}}{(r_p^{(k-1)/k} - 1)_{ic} (\% \, H_{ad})}\right]^{1/2} \tag{7-20}$$

For the constant weight flow process, the discharge-pressure Equations 7-11 and 7-12 must further be refined by the ratio of the adiabatic heads at the new volume flow and at the rated volume flow:

$$P_2 = P_1 \left\{ \left[\frac{H_{ad}}{(H_{ad})_r}\right] \left[\frac{(T_1)_r}{T_1}\right] \left[\frac{Z_r}{Z}\right] \left[\frac{MW}{MW_r}\right] \left[\frac{k}{k-1}\right]_r \left[\frac{k-1}{k}\right] \cdot [(r_p)_r^{[(k-1)/k]} - 1] + 1 \right\}^{k/(k-1)} \tag{7-22}$$

Equation 7-20 holds for any given set of inlet conditions. Therefore, the required corrections due to changes in molecular weight, inlet temperature, ratio of specific heats, or compressibility must be made prior to using Equation 7-20. If one wishes to compare the more general case of variable inlet conditions, Equation 7-23 must be used:

$$N_{req} = N_r \left[\frac{(H_{ad})_{req}}{(H_{ad})_r \times \% \, H_{ad}}\right]^{1/2} \tag{7-23}$$

Polytropic head can also be substituted for adiabatic head.

Recall that for the compressibility in all equations derived from the head equation, one should use:

Z_1 (inlet) for single-stage compressors

Z_a (average) for multi-stage compressors

In any case, the inlet volume flow is computed from the inlet compressibility.

These techniques were derived for a single-stage air compressor. Use of them on centrifugals having several impellers between flanges will yield useful qualitative results. Note that since volume ratio effects were ignored, the accuracy of these procedures will deteriorate in proportion to the number of impellers and the variation from rated of the inlet conditions. When using these techniques for multi-stage compressors, the techniques must be applied between each pair of flanges.

8

Centrifugal Compressor Rerates*

There are several reasons why one would want to rerate a centrifugal compressor: to increase plant capacity; to change the process pressure levels; to adapt to an entirely different process; and to take advantage of the latest technology for efficient compressor staging. Any of these may be accomplished by rerating existing compressors as opposed to purchasing new apparatus. A compressor rerate will usually cost no more than half the purchase price of a new machine. The lube system can normally be reused, and most of the installation costs can be saved, including those of piping and foundations. If the existing compressor has operated without trouble in the past, the chance of continued success of the rerated compressor may be better than that of a new unit, since past experience is a better tool than any computer.

An existing centrifugal compressor can be tested for rerate candidacy by investigating four major considerations:

1. *Capacity*. Will the size of the inlet nozzle accept the projected increase in flow?
2. *Pressure*. Can the casing accommodate the intended pressure levels? Can the compressor aerodynamically meet these levels?
3. *Power*. Will the driver combination handle the increased power requirement?

*Adapted by special permission from *Chemical Engineering*, Jan. 20, 1975, © 1975 by McGraw-Hill, Inc., New York, NY 10020; originally written for the CAGI Technical Article Program.

4. *Speed.* Will the compressor handle the required speed within critical speed limitations?

These four considerations will be reviewed individually.

Capacity

Normally, compressor manufacturers can provide impellers to handle any capacity that falls within the velocity limitations of their standard inlet nozzle sizes. It can therefore be reasonably assumed that the size of the inlet nozzle formulates the capacity limit of a compressor. It must be determined whether the existing inlet nozzle will pass the required volume flow with a reasonable pressure drop. Since the volume flow and pressure drop are both functions of the inlet velocity, the analysis can be reduced to a consideration of inlet velocities.

A good rule of thumb sets the maximum inlet velocity for reasonable pressure drop and flow distribution at 140 ft/sec (43 m/s) for air and lighter gases.

In order to prevent high inlet Mach numbers (ratio of inlet velocity to speed of sound of gas) and potential mal-distribution of flow into the first impeller, the maximum inlet velocity must be reduced for heavier gases and low-temperature applications.

This can best be handled by comparing the sonic velocity of the subject gas to the sonic velocity of air at 80°F (27°C). Consider Equations 5-12 and 5-13, presented here for convenience:

$$\theta = \sqrt{\frac{26.1\,MW}{k_1\,Z_1\,T_1}} \quad \text{in the English system} \tag{5-12}$$

$$\theta = \sqrt{\frac{14.5\,MW}{k_1\,Z_1\,T_1}} \quad \text{in the metric system} \tag{5-13}$$

By inverting these equations:

$$\theta^{-1} = \sqrt{\frac{k_1\,Z_1\,T_1}{26.1\,MW}} \quad \text{in the English system} \tag{8-1}$$

$$\theta^{-1} = \sqrt{\frac{k_1\,Z_1\,T_1}{14.5\,MW}} \quad \text{in the metric system} \tag{8-2}$$

The maximum acceptable nozzle velocity can then be calculated using Equations 8-3 and 8-4:

$$V_{max} = 140\sqrt{\frac{k_1 Z_1 T_1}{26.1\ MW}} \leq 140 \text{ ft/sec} \quad \text{English} \tag{8-3}$$

$$V_{max} = 43\sqrt{\frac{k_1 Z_1 T_1}{14.5\ MW}} \leq 43 \text{ m/s} \quad \text{metric} \tag{8-4}$$

Equations 8-5 and 8-6 determine the actual inlet velocity:

$$V_a = 3.06\ Q_1/d_n^2 \quad \text{in the English system} \tag{8-5}$$

$$V_a = 354\ Q_1/d_n^2 \quad \text{in the metric system} \tag{8-6}$$

where:
Q_1 is in ICFM (English) or m^3/h (metric)
d_n is in inches (English) or mm (metric)
V_a is in ft/sec (English) or m/s (metric)

Since velocity limitations set the maximum capacity, the maximum velocity of 140 ft/sec (43 m/s) can be limiting when rerating a compressor. With geometrically sound inlet piping, higher velocities may be found possible; therefore, these limits should not be considered absolute but rather a good engineering guide. The final recommendation lies with the manufacturer.

Some compressors have inlet nozzles other than the main inlet. Some have side-load nozzles, some have interstage cooling nozzles, some have a combination of both. The foregoing inlet velocity criteria can be used for all inlet nozzles, and all inlet nozzles should be checked for gas velocities.

Once it is determined that the compressor can pass the desired capacity, the feasibility of any required pressure level changes or changes in the polytropic head can be investigated.

Pressure

The compressor manufacturer, at the time of manufacture, hydrostatically tests the casing to 150% of the maximum expected operating pressure. This test is essentially a safety test for casing integrity. For any process, compressor aerodynamics may set the maximum operating pressure, or the process may set it by the existence of a relief valve in the system.

If the expected pressure levels of the new process exceed the nameplate maximum pressure, a new hydrostatic test should be considered. *Caution:* Sometimes, in a compressor made up of two or more sections (as with interstage cooling, side-loads, or extractions), the test engineer will test the compressor sectionally, with the low-pressure section at a lower test pressure than the high-pressure section. In such a case, both sections must be examined for the possibility of exceeding the maximum sectional safe operating pressure.

Next, the compressor's aerodynamic ability to deliver the required head must be investigated. This investigation can be made through the consideration of the polytropic head. The general form of the polytropic head equation is Equation 2-12:

$$H_p = ZRT_1 \frac{n}{n-1} [r_p^{(n-1)/n} - 1] \tag{2-12}$$

In order to calculate $n/(n - 1)$, a polytropic efficiency must be assumed. To reduce error, the investigation and feasibility analysis should be performed by ratioing new process requirements to existing compressor performance. Therefore, any error introduced by an inaccurate polytropic efficiency will be minimized. If the role of the polytropic efficiency in Equation 2-12 is analyzed, the error in the $n/(n - 1)$ multiplier will be found to counterbalance the error in the $(n - 1)/n$ exponent, thus considerably reducing error in the calculated polytropic head. Furthermore, errors in the new calculated polytropic head requirement will approximate those in the calculated head for existing performance. The ratio of the two will practically eliminate any errors. Therefore, while polytropic efficiencies will range from about 68% to 83%, a feasibility analysis can be conducted by assuming an average polytropic efficiency for all cases. A good number to use would be 76%.

Equation 2-12 can be used to calculate both the original polytropic head and the rerate polytropic head requirement. Witn the original rated speed known, the fan law, Equation 8-7, determines the approximate rerate speed:

$$N_{\text{rerate}} = N_{\text{original}} (H_{p_{\text{rerate}}}/H_{p_{\text{original}}})^{1/2} \tag{8-7}$$

The speed calculated from Equation 8-7 may turn out too high for safe operation (see the considerations of rotational speed later in this chapter). However, the manufacturer can sometimes obtain increased polytropic head by adding an impeller or replacing the existing impellers with ones of lower backward lean. (The higher the backward lean, the lower the polytropic head output but the higher the head rise to surge, and therefore the greater the stability.) The manufacturer will ultimately have to determine the feasibility of obtaining the required head.

A quick estimate can be made of the maximum head capability of an existing compressor. If the compressor contains a "blank" stage (stage space with no impeller), Equation 8-8 will yield the approximate maximum head capability with no change in speed:

$$H_{P_{max}} = H_{P_{original}}[(e + f)/e] \qquad (8\text{-}8)$$

where:

e = number of impellers in original rotor
f = number of blank stages in original rotor

The fan-law, along with Equation 8-8, can then be applied to determine the approximate polytropic head capability of the compressor. This procedure should yield a polytropic head within 10% of the maximum compressor capability.

Note that sometimes the compressor manufacturer may decrease the diameter of one or more impellers. This practice reduces the polytropic head output. The purpose of trimming may be to satisfy intermediate process pressure levels, such as in the case of side-load compressors, or to satisfy head requirements on fixed-speed applications such as direct motor drives or gas-turbine drives. In any case, if the compressor has trimmed impellers, this may also serve as a vehicle for increased head production.

The foregoing procedure can be used to estimate the required polytropic head for compressors with side-loads, extractions, or interstage cooling. For a feasibility analysis of a multi-section compressor, the total head requirements should be calculated as the sum of the individual sectional head requirements for both the original and rerate conditions. Therefore, for multi-section compressors, determine the head for section 1 based on operating parameters for section 1 (inlet temperature, inlet pressure, discharge pressure), the head for section 2 based on the operating parameters for section 2, and so on. The sum of all sectional head requirements yields the total head requirement for the rerate. The total head requirement for the original conditions can be determined in a similar manner and the approximate rerate speed derived from Equation 8-7 as before. This approach works well; however, depending on process requirements, the head requirement in one section may change more than the head requirement in another. Thus, in the final analysis, some allowances must be made for variations in interstage pressure levels.

A change in both mass flow and polytropic head will affect the power requirement. The effects of the process change on the driver must be considered next.

Compressor Power

The change in power requirements is a major consideration in a rerate feasibility analysis. This is especially true in the case of motor-driven compressors, since oversized motors are not usually purchased with the original equipment. However, the manufacturer can usually rebuild turbines to provide greater power capacity. One can usually obtain greater power capacity from gears, also. Motor drivers may require buying a new motor, which may lead, in some cases, to foundation problems. In any case, the driver manufacturer will have to be consulted regarding increased power output capabilities.

The compressor will require power approximately proportional to the product of the change in weight flow and the change in head. This can be seen by reviewing the power equation, Equation 8-9:

$$PWR = \frac{mH_p}{C_1\,\eta_p} + L_m \qquad (8\text{-}9)$$

Therefore, a 10% increase in mass flow coupled with a 10% increase in polytropic head will result in an increase of approximately 21% in power requirement (1.1 × 1.1 = 1.21). There will almost inevitably be some change in compressor efficiency when rerating a compressor. On the other hand, the mechanical losses of the compressor are a function of speed and not duty. Therefore, the mechanical losses will remain almost constant from original to rerate conditions. However, the effects of these two factors will be minimal, and rerate power requirements can be estimated from Equation 8-10:

$$PWR_{\text{rerate}} = PWR_{\text{original}} \left[\frac{H_{P\text{rerate}}}{H_{P\text{original}}}\right]\left[\frac{m_{\text{rerate}}}{m_{\text{original}}}\right] \qquad (8\text{-}10)$$

If gears form part of the drive train, approximately 2% must be allowed for gear losses when reviewing driver capabilities.

Also, at least 10% extra power capability should be considered in the driver in order to take care of potential overload conditions (i.e. operating near stonewall, where power requirements are higher).

If the driver cannot deliver the required power to the compressor, either a new driver must be purchased or the scheduled process change relaxed until the required mass flow and polytropic head match driver capabilities. In cases where driver considerations or power requirements are critical, power considerations may set the feasible limit for compressor reratability.

Speed

The final consideration is that of rotational speed. Two major criteria must be satisfied:

1. The rotational speed must be slow enough to not overstress the impellers.
2. The rotational speed should fall within certain limits of the compressor's first and second critical speeds, as specified by the American Petroleum Institute (API Standard 617).

The first criterion can best be viewed from the standpoint of mechanical tip speed limitations. The mechanical tip speed can be determined by using Equation 6-1 or 6-2:

$$u = N \pi d/720 \quad \text{in the English system} \tag{6-1}$$

$$u = N \pi d/(6 \times 10^4) \quad \text{in the metric system} \tag{6-2}$$

Mechanical tip speed limits vary with manufacturer, size, impeller material, and type of construction. A typical limit on mechanical tip speed is 900 ft/sec (275 m/s). When using the fan law, the rotational speed that results in a tip speed of 900 ft/sec (275 m/s) for the largest impeller should be considered the upper limit (providing that this speed falls within API critical speed limitations).

One will not know what effect the rerate hardware will have on the compressor critical speeds when conducting the feasibility analysis. Since the bearing centerline distance is fixed, and since most rerates can be accomplished without a change of bearings, critical speeds will not generally be affected by the rerate. Therefore, for a rerate feasibility analysis, compare the rerate rotational speed to the original critical speed.

EXAMPLE 8-1

Consider a straight-through centrifugal compressor on a dry-air process, having the following nameplate data:

$Q = 18{,}000 \text{ m}^3/\text{h}$

$T_1 = 30°\text{C}$

$P_1 = 100 \text{ kPa}$

P_2 = 400 kPa

PWR = 1,300 kW

N = 8,100 RPM

N_{max} = 8,500 RPM

N_{c1} = 4,800 RPM (N_{c1} is the first critical speed)

MW = 28.97

k_a = 1.4

Z_a = 1.0

P_{2max} = 475 kPa.

The desired rerate is an increase of the inlet capacity to 21,000 m^3/h, and an increase of the discharge pressure to 450 kPa, with all other inlet conditions unchanged.

An investigation of the data files on the compressor reveals that the second critical speed, N_{c2}, is 10,800 RPM, and that the largest wheel diameter is 560 mm. Also, thc cross-sectional drawing of the compressor indicates that the inlet nozzle diameter is 508 mm.

1. Compute the maximum feasible inlet velocity using Equation 8-4 (for simplicity, use k_a and Z_a for k_1 and Z_1, respectively):

$$V_{max} = 43\sqrt{\frac{k_1 Z_1 T_1}{14.5\ MW}} = 43\sqrt{\frac{(1.4)(1{:}0)(273\ +\ 30)}{(14.5)(28.97)}} = 43.2 \text{ m/s} > 43.0$$

   i.e. V_{max} = 43 m/s.

2. Using Equation 8-6, compute the inlet velocity based on the new inlet volume flow:

$$V_a = 354\ Q_1/d_n^2 = 354\left[\frac{21{,}000}{(508)^2}\right] = 28.8 \text{ m/s} < 43 \text{ m/s}$$

   Since V_a is less than V_{max}, the proposed capacity is feasible.

3. Since the nameplate maximum discharge pressure is 475 kPa, the compressor would not have to be hydrostatically retested, provided that the process will not allow the value of 475 kPa to be exceeded.

4. From Equation 2-16, calculate the polytropic head for both the original and the rerate conditions:

Original:

$$\frac{n}{n-1} = \frac{k}{k-1}\eta_p = \frac{1.4}{1.4-1}(0.76) = 2.66$$

$$H_p = \frac{Z_a}{1{,}000}\frac{8{,}314}{MW}T_1\frac{n}{n-1}[r_p^{(n-1)/n} - 1]$$

$$= \frac{1}{1{,}000}\left(\frac{8{,}314}{28.97}\right)(303)(2.66)\left[\left(\frac{400}{100}\right)^{1/2.66} - 1\right]$$

$$= 158.2 \text{ kN·m/kg}$$

Rerate:

$$H_p = \frac{1}{1{,}000}\left(\frac{8{,}314}{28.97}\right)(303)(2.66)\left[\left(\frac{450}{100}\right)^{1/2.66} - 1\right]$$

$$= 175.8 \text{ kN·m/kg}$$

Determine the new required speed from the fan law, Equation 8-7:

$$N_{\text{rerate}} = N_{\text{original}}\sqrt{H_{p_{\text{rerate}}}/H_{p_{\text{original}}}}$$

$$= 8{,}100\sqrt{175.8/158.2} = 8{,}540 \text{ RPM}$$

The largest wheel diameter is 560 mm; therefore, from Equation 6-2:

$$u = \frac{N\,\pi\,d}{6\times10^4} = \frac{(8{,}540)\,\pi\,(560)}{6\times10^4} = 250 \text{ m/s} < 275 \text{ m/s}$$

The new rotational speed results in a satisfactory mechanical tip speed. API states that the second critical speed must be 20% above the highest operating speed. Assuming that the new rerate speed is the highest for the new process, the second critical speed must be at least:

$$(1.2)(8{,}540) = 10{,}250 \text{ RPM}$$

The second critical speed (10,800 RPM) is higher than the required; therefore, the rotational speed is feasible.

5. The total increase in power due to the changes in both mass flow and polytropic head can now be estimated. Since the rated inlet conditions have not changed, the increase in mass flow will be proportional to the increase in

volume flow; therefore, the power requirement due to the change in volume flow will increase by the same proportion:

$$\frac{PWR_{\text{rerate}}}{PWR_{\text{original}}} \propto \frac{m_{\text{rerate}}}{m_{\text{original}}} \propto \frac{Q_{\text{rerate}}}{Q_{\text{original}}}$$

The new power requirement, due to changes in both head and flow, will be:

$$PWR_{\text{rerate}} = PWR_{\text{original}}\left[\frac{H_{p\text{rerate}}}{H_{p\text{original}}}\right]\left[\frac{Q_{\text{rerate}}}{Q_{\text{original}}}\right]$$

$$= (1{,}300)\left(\frac{175.8}{158.2}\right)\left(\frac{21{,}000}{18{,}000}\right) = 1{,}685 \text{ kW}$$

The driver must therefore be capable of:

$$(1.1)(1685) = 1{,}855 \text{ kW}$$

plus 2% if a gear is involved.

Since the inlet velocity, maximum operating pressure, and required rotational speed are within satisfactory limits, the rerate is feasible, provided the driver can meet the necessary power requirements.

Other Considerations

The foregoing has reviewed the four major considerations of a rerate feasibility analysis: capacity, pressure, power, and speed. A centrifugal compressor passing these tests will be a good candidate for rerating. The compressor hardware is normally the most expensive investment in a rerate; therefore, the decision to procede with the rerate will probably be made primarily on the cost of revamping the compressor. However, there are some other factors one should be aware of, or that the manufacturer, at least, should be instructed to consider:

1. Shaft torque
2. Couplings
3. Lube system

Unless the material and heat treatment of the existing shaft are known, one will not be able to check whether the shaft can carry the torque for the rerate. However, the new torque can be compared to the original torque to gain a feeling for the amount of extra torque the shaft will have to carry. The torque is proportional to the ratio of the shaft power to the speed of rotation, as shown in Equation 8-11:

$$\tau \propto \frac{PWR_s}{N} \tag{8-11}$$

where:
τ = torque

This ratio could be computed for both the rerate and original conditions and compared. In any event, the manufacturer should be requested to check the shaft for torque-carrying capability and report his findings. Should the shaft be incapable of transmitting the required torque, a new shaft will be required either of higher strength material or with a larger shaft end.

Couplings are another consideration for a rerate. Couplings can normally carry higher torque than the shaft to which they are coupled. Therefore, for torque considerations, chances are if the shaft end is satisfactory for the rerate, the coupling will be also. However, if the shaft end is increased, a new coupling will be required, with a hub matched to the new shaft end. Moreover, if any rotating equipment in the string is changed, such as the gear or prime driver, the coupling mating with that equipment will probably have to be replaced.

Torque is not the only consideration for the couplings. The couplings must also "tune" the rotating string of equipment so that no torsional critical speeds interfere with the desired speed range. Therefore, the reader should always request a torsional critical speed analysis with a compressor rerate. In cases where a gear is present in the string, or where the string is composed of two or more compressor bodies, a torsional analysis is highly advisable.

The final recommendation for coupling replacement rests with the manufacturer. However, the manufacturer should be requested to review the existing couplings for torque and proper fit, perform a torsional analysis of the rerated rotating string to determine torsional adequacy, and formally report the results. (The manufacturer may not be willing to perform the torsional analysis at the time of rerate study; he may wish to wait until he receives an order. However, this is not detrimental, since the manufacturer will probably be able to find couplings that will result in a torsionally sound system.)

The last consideration is the lube system. Since the bearings and seals in the compressor are not normally changed for a rerate, and since the new speed requirements will normally be within the speed range considered by the manufacturer at the time of original sale, lube requirements for the compressor do not normally increase as a result of a rerate. However, changes in the drive train, such as the gear, can alter lube requirements. While requirements for changes in the lube system are not common, they should certainly be checked as part of the rerate study. When lube-system modifications are required, they can usually be handled by a change in pumps, coolers, and filters. If any more of the lube system requires modification, it would hardly be worth the expense to modify, and a new lube system would be in order. Not much can be done to test the feasibility of reusing the lube system; however, without driver change, lube system modifications are not generally necessary. The final check on the lube system rests with the manufacturer. The reader should request that the manufacturer check the lube system and formally report the results of his investigation.

Appendix A
The Polytropic Head Equation and the Relationship Between *k*-Value and *n*-Value

Polytropic Head Equation

For a steady-state, steady-flow process, the polytropic head is given by the following integral relationship:

$$H_p = \int_1^2 v dP$$

The polytropic process is of the form:

$$Pv^n = \text{constant}$$

Then:

$$Pv^n = \text{constant} = C^n \rightarrow v = \frac{C}{P^{1/n}}$$

Substituting the preceding into the integral relationship for the polytropic head:

$$H_p = \int_1^2 \frac{CdP}{P^{1/n}} = \frac{C}{1 - \dfrac{1}{n}} P^{1-(1/n)} \Bigg|_1^2$$

$$= C \frac{n}{n-1} P^{(n-1)/n} \Big|_1^2$$

$$= C \frac{n}{n-1} (P_2^{(n-1)/n} - P_1^{(n-1)/n})$$

$$= C \frac{n}{n-1} P_1^{(n-1)/n} \left[\left(\frac{P_2}{P_1} \right)^{(n-1)/n} - 1 \right]$$

But:

$$Pv^n = C^n$$

$$C = (Pv^n)^{1/n} = P_1^{1/n} v_1$$

Referring back to the last relationship for H_p, consider $CP_1^{(n-1)/n}$:

$$CP_1^{(n-1)/n} = P_1^{1/n} v_1 P_1^{(n-1)/n} = P^{[1/n]+[(n-1)/n]} v_1 = P_1 v_1$$

From the ideal gas law:

$$P_1 v_1 = Z_1 R T_1$$

Substituting the ideal gas law:

$$H_p = C \frac{n}{n-1} P_1^{(n-1)/n} \left[\left(\frac{P_2}{P_1} \right)^{(n-1)/n} - 1 \right]$$

$$= P_1 v_1 \frac{n}{n-1} \left[\left(\frac{P_2}{P_1} \right)^{(n-1)/n} - 1 \right]$$

$$= Z_1 R T_1 \frac{n}{n-1} \left[\left(\frac{P_2}{P_1} \right)^{(n-1)/n} - 1 \right]$$

Since P_2/P_1 is defined as the pressure ratio, r_p:

$$H_p = Z_1 R T_1 \frac{n}{n-1} [r_p^{(n-1)/n} - 1]$$

Relationship Between *k*-Value and *n*-Value

The efficiency of a polytropic compression process is defined as the ratio of the work output to the work input. The work output per mass of fluid is the polytropic head. For an ideal gas with constant specific heats, the polytropic head is:

$$H_p = RT_1 \frac{n}{n-1} [r_p^{(n-1)/n} - 1]$$

The work input per mass of fluid is the change in enthalpy of the gas:

$$w = \Delta h = c_p \Delta T = c_p(T_2 - T_1)$$

$$T_2 - T_1 = T_1\left(\frac{T_2}{T_1} - 1\right) = T_1(r_t - 1) = T_1(r_p^{(n-1)/n} - 1)$$

$$w = c_p T_1 (r_p^{(n-1)/n} - 1)$$

But:

$$c_p = R\frac{k}{k-1}$$

Therefore:

$$w = RT_1 \frac{k}{k-1} (r_p^{(n-1)/n} - 1)$$

The polytropic efficiency is defined as:

$$\eta_p = \frac{H_p}{W}$$

Then:

$$\eta_p = \frac{RT_1 \frac{n}{n-1} [r_p^{(n-1)/n} - 1]}{RT_1 \frac{k}{k-1} [r_p^{(n-1)/n} - 1]}$$

$$= \frac{\frac{n}{n-1}}{\frac{k}{k-1}}$$

And:

$$\frac{n}{n-1} = \frac{k}{k-1}\eta_p$$

Appendix B
Summary of Centrifugal Compressor Equations and Estimating Data

The following figures and tables are included in this section:

Typical centrifugal compressor frame data are found on page 54, Table 5-1.
An efficiency conversion chart may be found on page 98, Figure 5-11.
A graph for determining maximum polytropic head per stage (English system) is found on page 61, Figure 5-2.
A graph for determining maximum polytropic head per stage (metric system) is found on page 62, Figure 5-3.

Table B-1
Centrifugal Compressor Equations

Gas Properties

$$P_r = \frac{P}{P_c} \qquad T_r = \frac{T}{T_c} \qquad Z_a = \frac{Z_1 + Z_2}{2}$$

$$k_a = \frac{k_1 + k_2}{2} \qquad k = \frac{c_p}{c_p - 1.986} \text{ (English);} \qquad k = \frac{c_p}{c_p - 8.314} \text{ (metric)}$$

$$MW_{\text{mix}} = yMW_w + (1 - y)MW_{dg} \qquad \frac{n}{n-1} = \frac{k}{k-1}\eta_p$$

Compressor Ratio Equations

$$r_p = r_t^{n/(n-1)} \qquad r_t = r_p^{(n-1)/n} \qquad r_v = r_p^{1/n}$$

$$r_p = r_v^n \qquad r_t = r_v^{n-1} \qquad r_v = r_t^{1/(n-1)}$$

Mass Flow

$$m_{\text{total}} = m_{dg}\left[1 + \frac{MW_w P_w}{MW_{dg} P_{dg}}\right]$$

$$m = SVF\left[\frac{144\, P_{\text{std}} MW_{\text{std}}}{1{,}545\, Z_{\text{std}}\, T_{\text{std}}}\right] \text{ (English);} \qquad m - SVF\left[\frac{MW_{\text{std}} P_{\text{std}} \times 10^3}{8{,}314\, Z_{\text{std}}\, T_{\text{std}}}\right] \text{ (metric)}$$

Temperature

$$T_2 = r_t T_1 = r_p^{(n-1)/n} T_1; \qquad T_2 = \frac{T_1[r_p^{(k-1)/k} - 1]}{\eta_{ad}} + T_1; \qquad T_{\text{mix}} = \frac{m_1 c_{p1} T_1 + m_2 c_{p2} T_2}{m_1 c_{p1} + m_2 c_{p2}}$$

Speed

$$N = N_{\text{nom}}\sqrt{\frac{H_p}{H_{p\text{nom}} \times \text{no. stages}}}$$

Fan Laws

$$Q \propto N \qquad H \propto N^2 \qquad PWR \propto N^3$$

Speed-of-Sound Index

English	***Metric***
$\theta = \sqrt{\dfrac{26.1\, MW}{k_1 Z_1 T_1}}$	$\theta = \sqrt{\dfrac{14.5\, MW}{k_1 Z_1 T_1}}$
$\theta = \sqrt{\dfrac{280}{k_1 P_1 v_1}}$	$\theta = \sqrt{\dfrac{121}{k_1 P_1 v_1}}$

Table B-1 (continued)

Enthalpy

$$h_2 = \frac{h_{2ad} - h_1}{\eta_{ad}} + h_1$$

Volume Flow

$$Q_1 = m \frac{1{,}545\, Z_1 T_1}{144\, MW\, P_1} \text{ (English)}; \quad Q_1 = m \frac{8{,}314\, Z_1 T_1}{MW\, P_1 \times 10^3} \text{ (metric)}$$

$$Q_1 = SVF \left(\frac{Z_1}{Z_{std}}\right)\left(\frac{T_1}{T_{std}}\right)\left(\frac{P_{std}}{P_1}\right)\left(\frac{MW_{std}}{MW}\right)$$

Head

English

$$H_p = Z_a \frac{1{,}545}{MW} T_1 \frac{n}{n-1} [r_p^{(n-1)/n} - 1]$$

$$H_{ad} = Z_a \frac{1{,}545}{MW} T_1 \frac{k}{k-1} [r_p^{(k-1)/k} - 1]$$

$$H_{ad} = 778(h_{2ad} - h_1)$$

$$H_p = H_{ad} \frac{\eta_p}{\eta_{ad}}$$

Metric

$$H_p = \frac{Z_a}{1{,}000} \frac{8{,}314}{MW} T_1 \frac{n}{n-1} [r_p^{(n-1)/n} - 1]$$

$$H_{ad} = \frac{Z_a}{1{,}000} \frac{8{,}314}{MW} T_1 \frac{k}{k-1} [r_p^{(k-1)/k} - 1]$$

$$H_{ad} = (h_{2ad} - h_1)$$

$$H_p = H_{ad} \left(\frac{\eta_p}{\eta_{ad}}\right)$$

Power

English

$$PWR_g = \frac{mH_p}{33{,}000\, \eta_p}$$

$$PWR_g = \frac{mH_{ad}}{33{,}000\, \eta_{ad}}$$

$$PWR_s = PWR_g + \text{mechanical losses}$$

Metric

$$PWR_g = \frac{mH_p}{3{,}600\, \eta_p}$$

$$PWR_g = \frac{mH_{ad}}{3{,}600\, \eta_{ad}}$$

$$PWR_s = PWR_g + \text{mechanical losses}$$

Sectional Pressure Ratio

$$r_{p\text{sect}} = \sqrt[b]{r_p}$$

Number of Stages

$$\text{no. stages} = \frac{H_p}{\text{max. } H_p/\text{stage}}$$

Table B-1 (Continued)

Units for Previous Equations

Parameter	*English*	*Metric*
P	psia	kPa
T	°R	K
v	ft^3/lbm	m^3/kg
MW	lbm/lbm-mol	kg/kmol
H	ft-lbf/lbm	kN·m/kg
m	lbm/min	kg/h
PWR	hp	kW
c_p	Btu/lbm-mol-°R	kJ/kmol·K

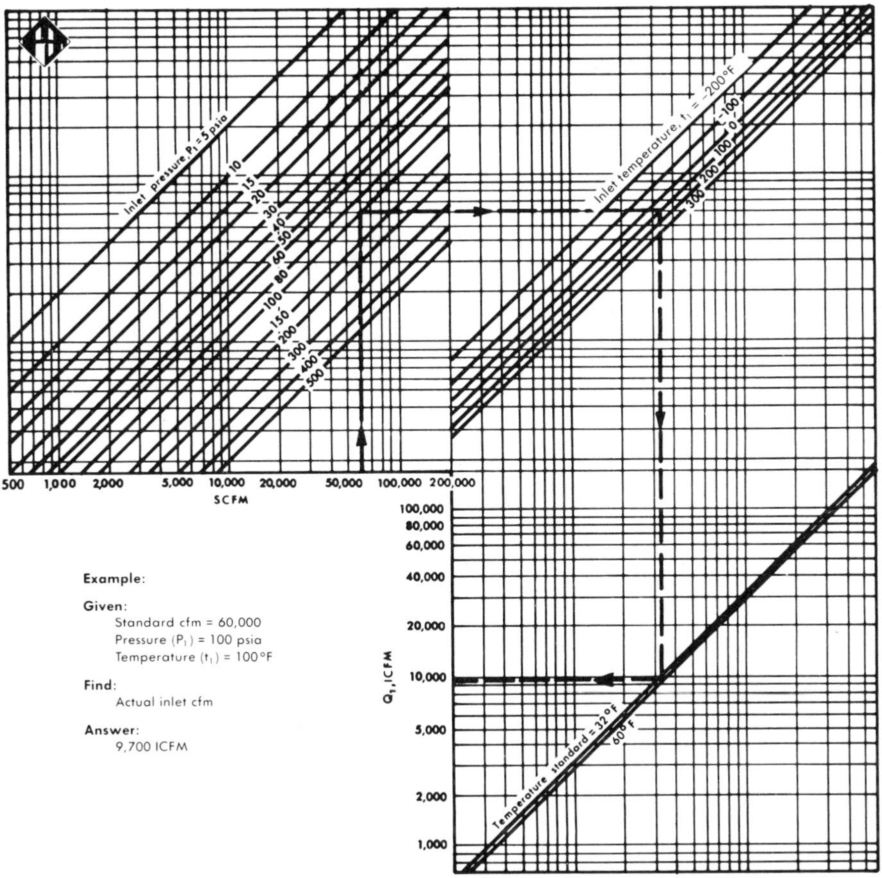

Figure B-1. Nomograph for converting standard volume flow to inlet volume flow—English system. Reprinted by permission of Gas Processors Suppliers Association *Engineering Data Book,* Ninth Edition, 1972, 5th Revision, 1981.

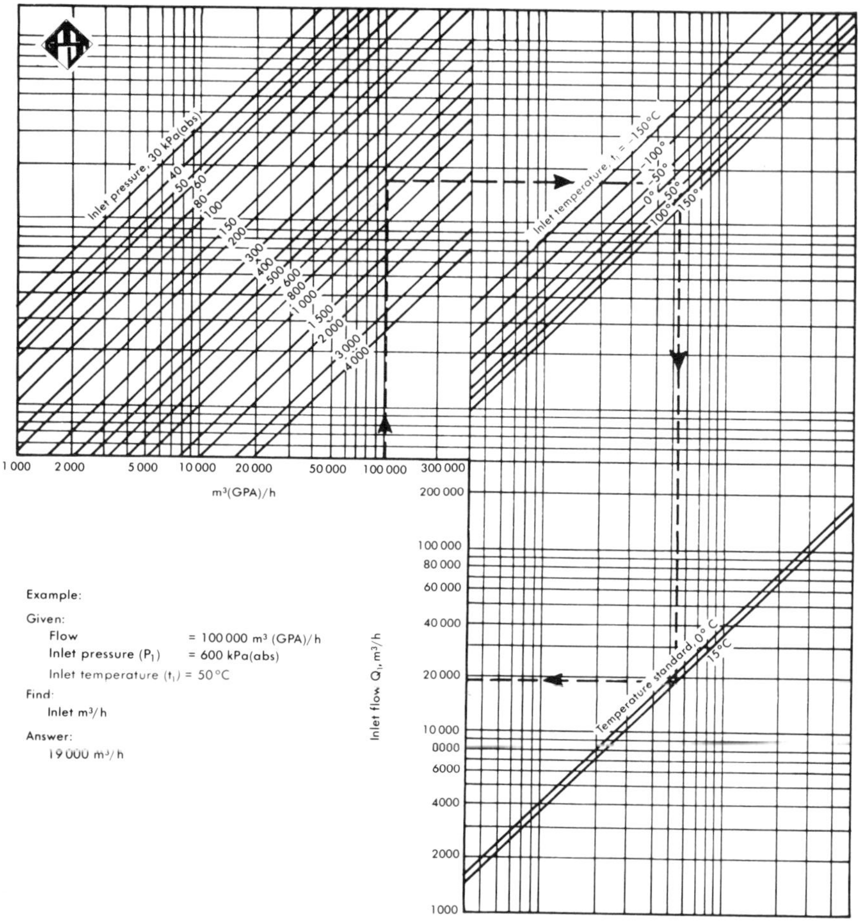

Figure B-2. Nomograph for converting standard volume flow to inlet volume flow—metric system. Reprinted by permission of Gas Processors Suppliers Association *SI Engineering Data Book,* 1980.

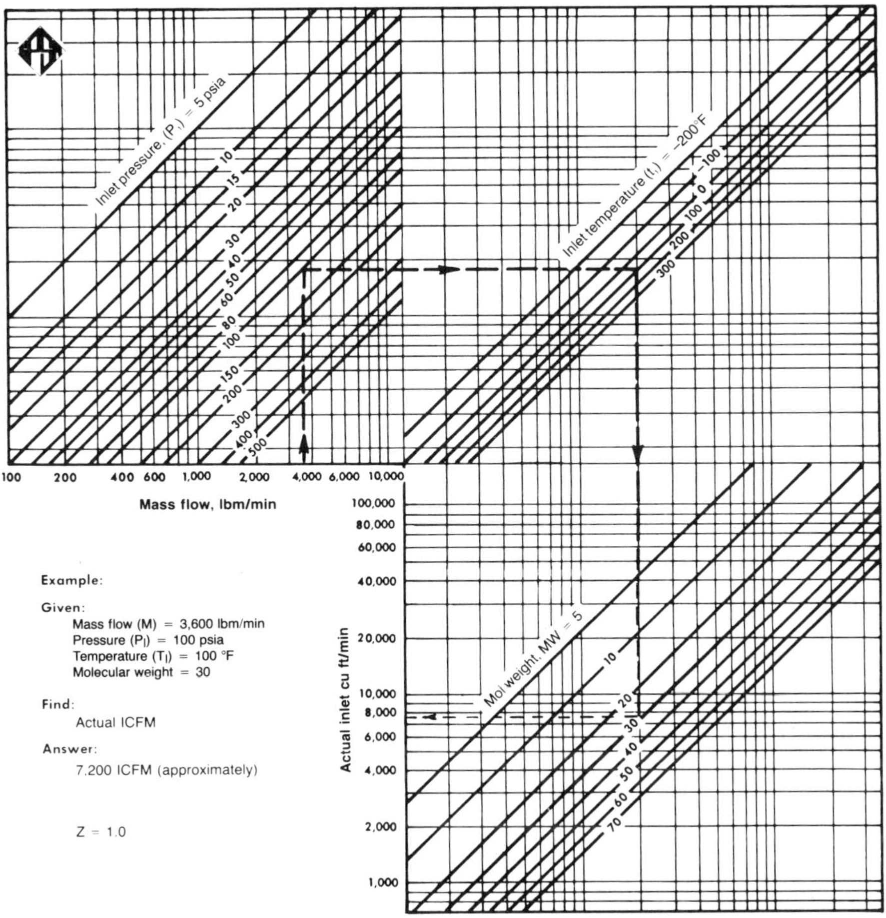

Figure B-3. Nomograph for converting mass flow to inlet volume flow—English system. Reprinted by permission of Gas Processors Suppliers Association *Engineering Data Book,* Ninth Edition, 1972, 5th Revision, 1981.

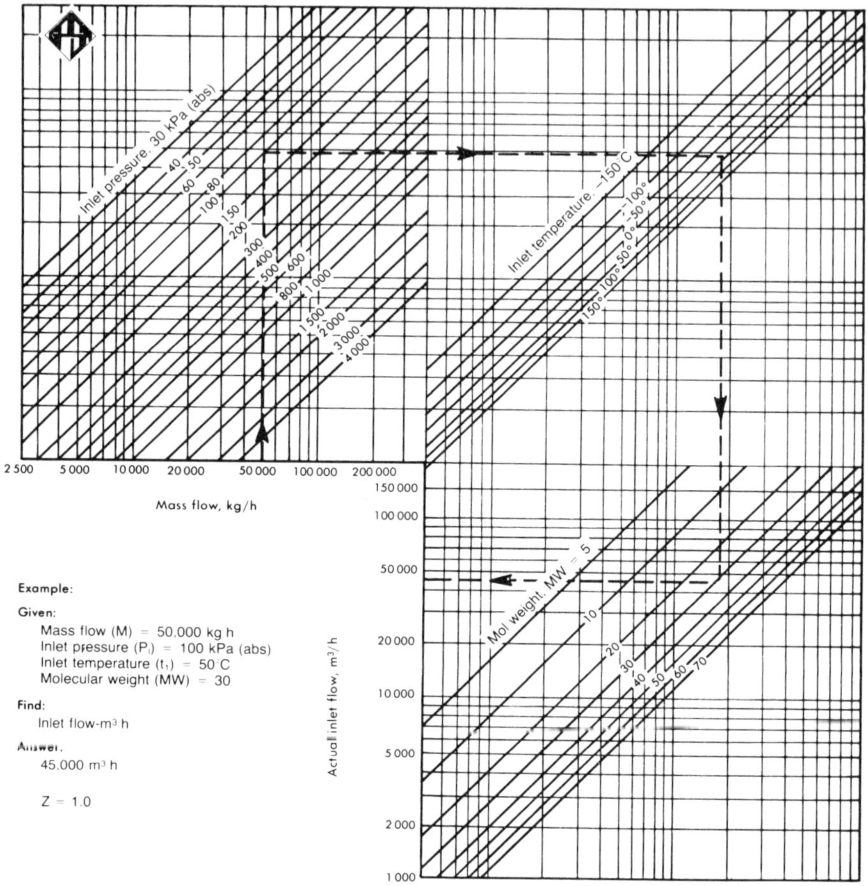

Figure B-4. Nomograph for converting mass flow to inlet volume flow—metric system. Reprinted by permission of Gas Processors Suppliers Association *SI Engineering Data Book*, 1980.

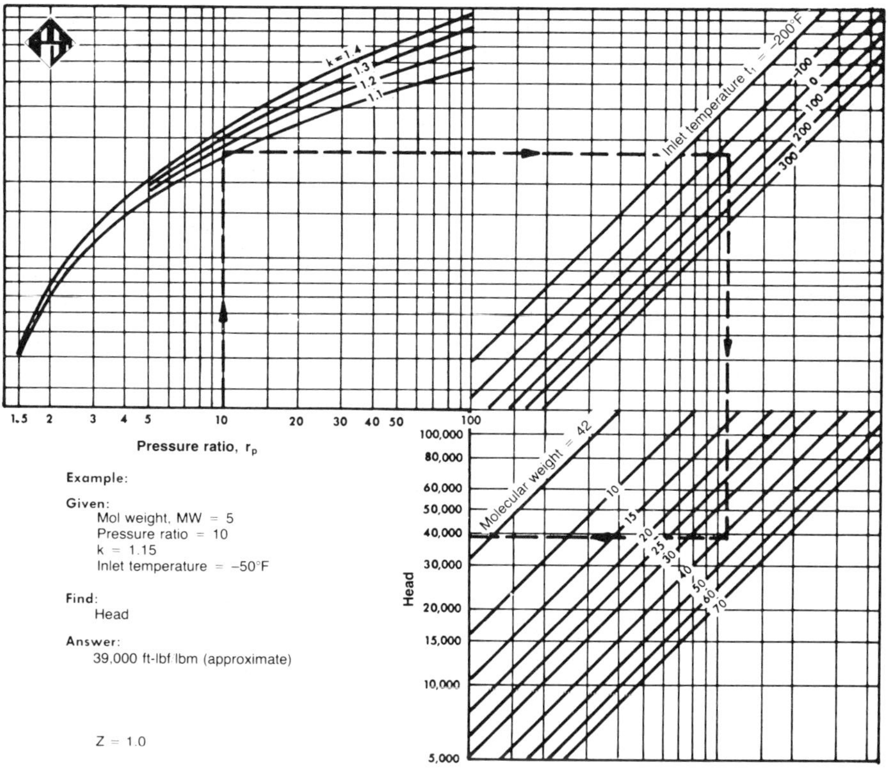

Figure B-5. Nomograph for converting presure ratio to adiabatic head—English system. Reprinted by permission of Gas Processors Suppliers Association *Engineering Data Book,* Ninth Edition, 1972, 5th Revision, 1981.

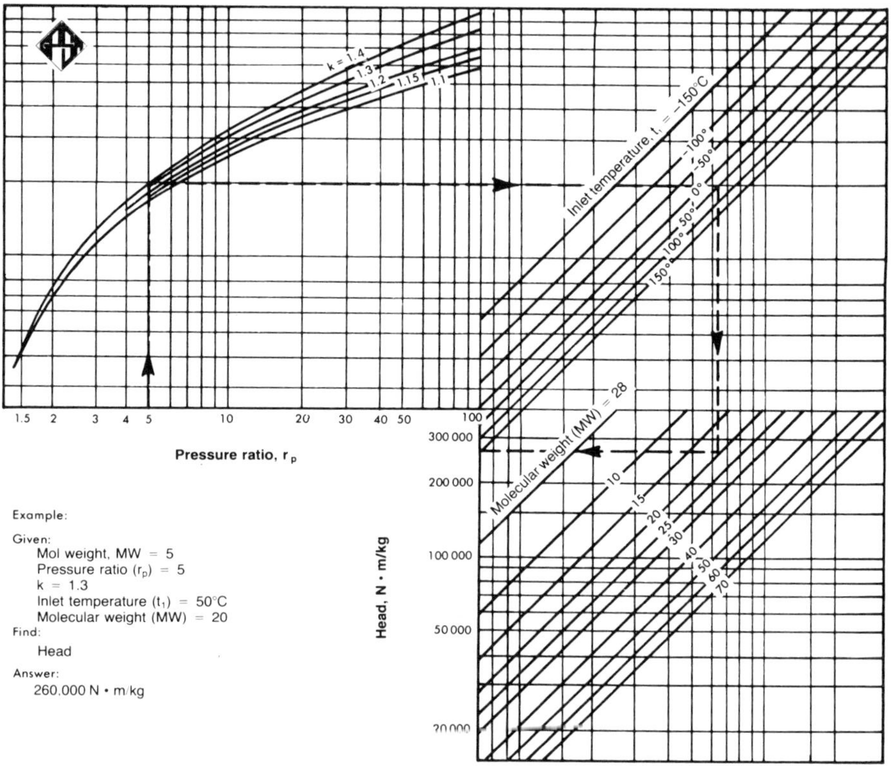

Figure B-6. Nomograph for converting pressure ratio to adiabatic head—metric system. Reprinted by permission of Gas Processors Suppliers Association *SI Engineering Data Book,* 1980.

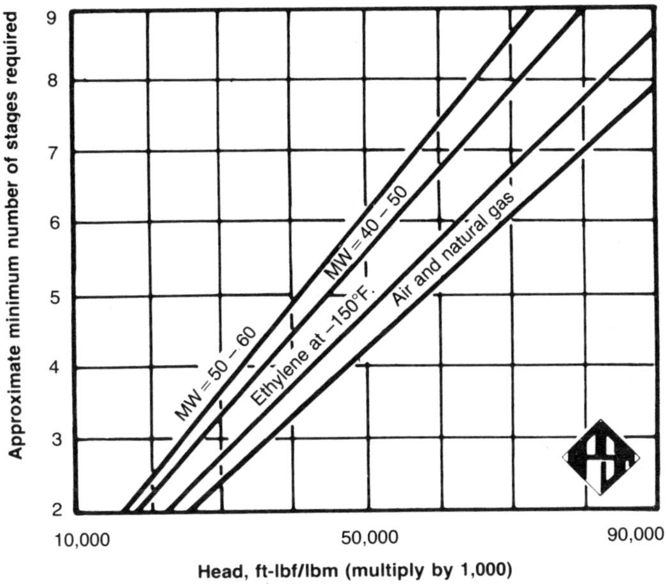

Figure B-7. Quick estimate for the required number of centrifugal compressor stages—English system. Reprinted by permission of Gas Processors Suppliers Association *Engineering Data Book,* Ninth Edition, 1972, 5th Revision, 1981.

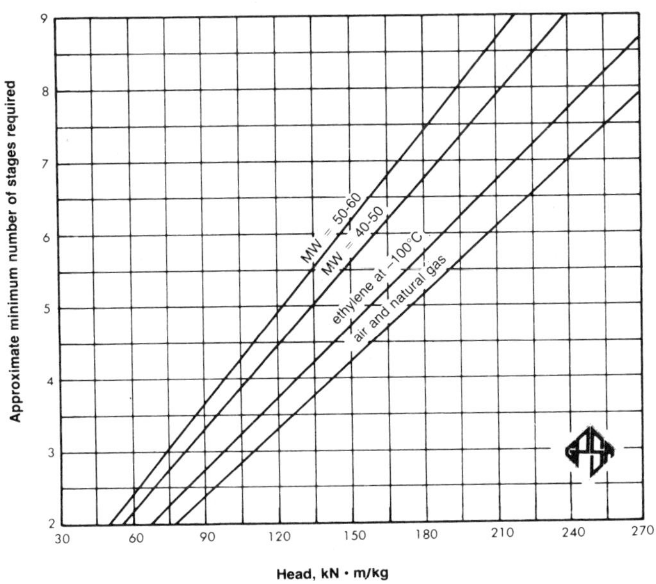

Figure B-8. Quick estimate for the required number of centrifugal compressor stages—metric system. Reprinted by permission of Gas Processors Suppliers Association *SI Engineering Data Book,* 1980.

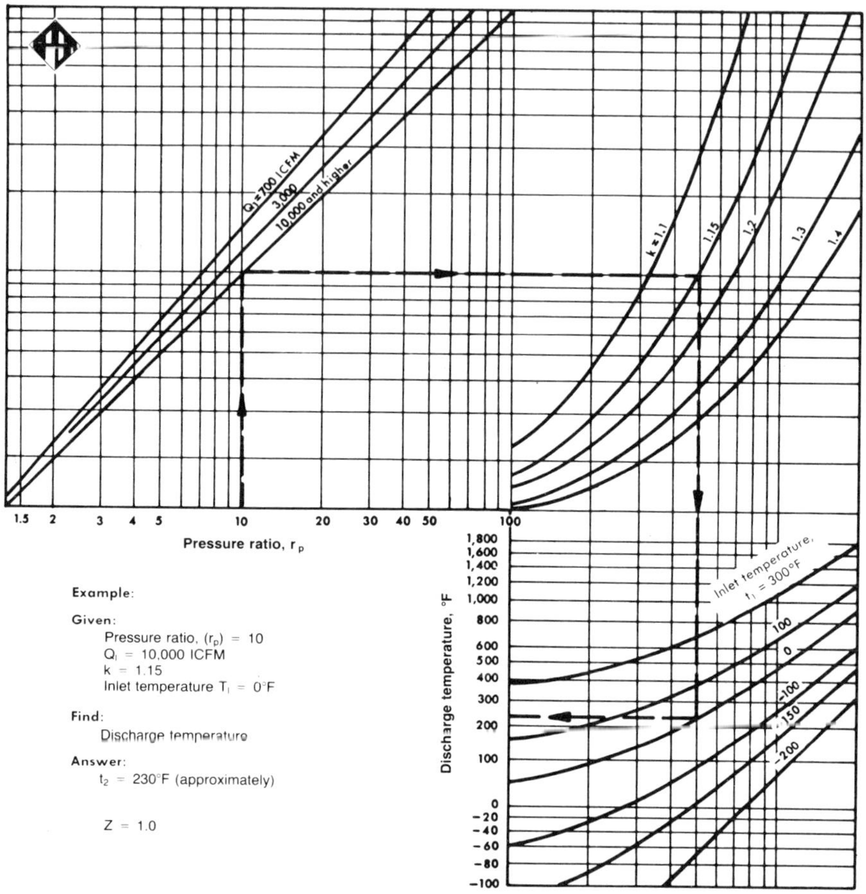

Figure B-9. Nomograph for estimating discharge temperature—English system. Reprinted by permission of Gas Processors Suppliers Association *Engineering Data Book,* Ninth Edition, 1972, 5th Revision, 1981.

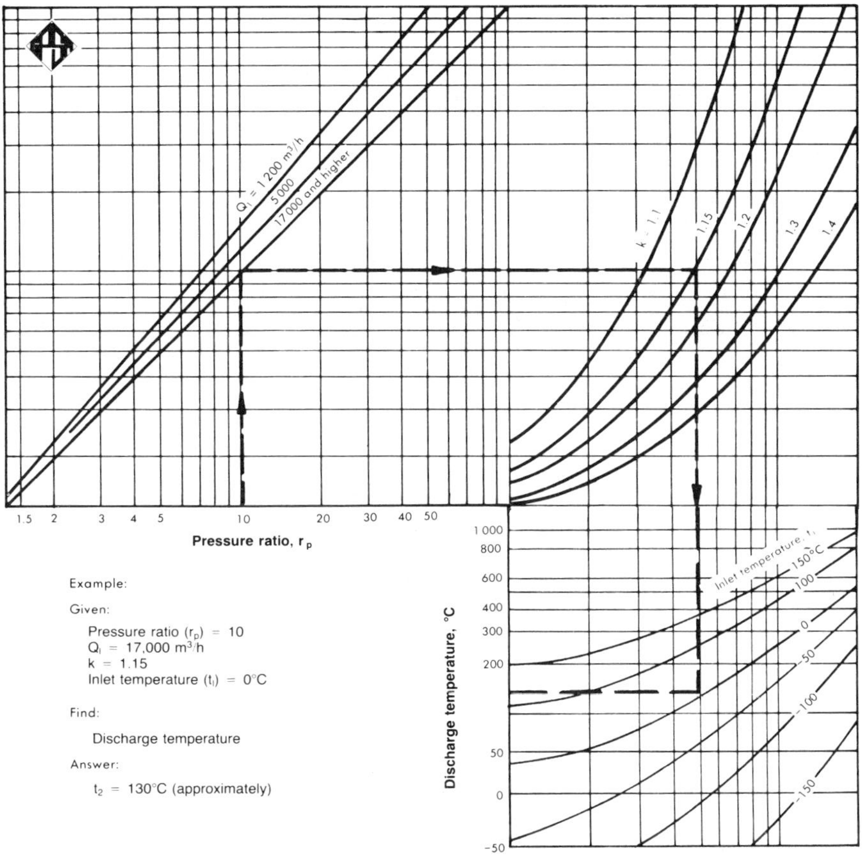

Figure B-10. Nomograph for estimating discharge temperature—metric system. Reprinted by permission of Gas Processors Suppliers Association *SI Engineering Data Book, 1980.*

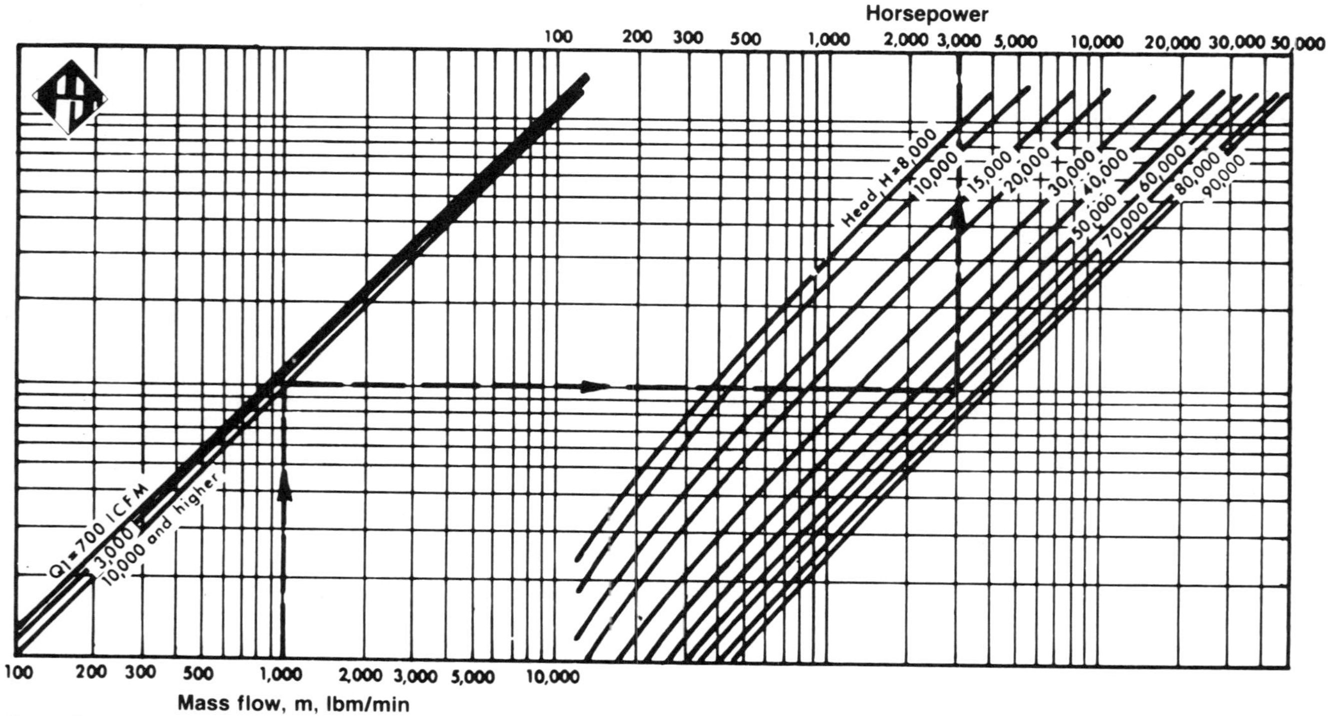

Example:

Given:

Mass flow, m = 1,000 lbm min
Head = 70,000 ft-lbf lbm

Find:

Gas horsepower

Answer:

Ghp = 3,000

Figure B-11. Nomograph for estimating power requirements—English system. Reprinted by permission of Gas Processors Suppliers Association *Engineering Data Book,* Ninth Edition 1972, 5th Revision, 1981.

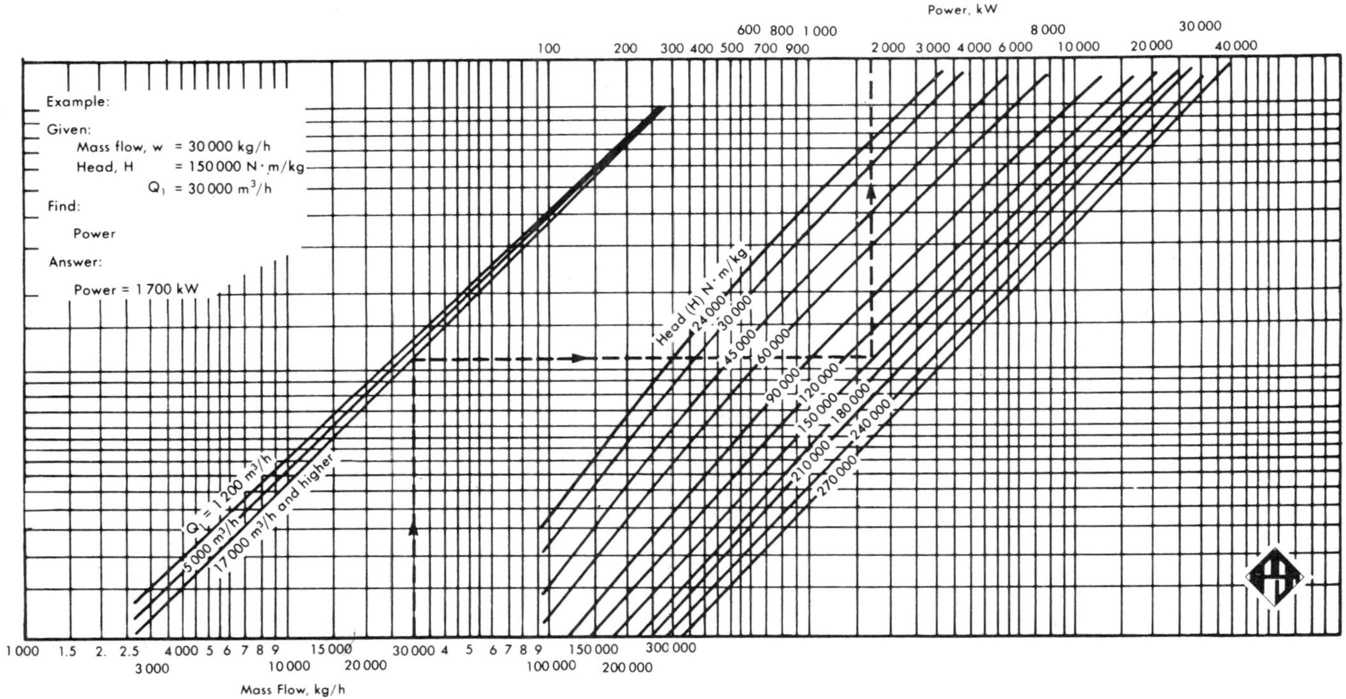

Figure B-12. Nomograph for estimating power requirements—metric system. Reprinted by permission of Gas Processors Suppliers Association *SI Engineering Data Book*, 1980.

Appendix C
Properties of Common Gases

The following figures and tables are included in this section:

Table C-1
Gas Properties*

Compound	Formula	Hydrocarbon Reference Symbol	Molecular Weight	Critical Constants: English Pressure, psia	Critical Constants: English Temperature, °R	Critical Constants: Metric Pressure, kPa	Critical Constants: Metric Temperature, K	Specific Heat, c_p: English Btu/lbm-mol-°R @ 60°F, 14.696 psia	Specific Heat, c_p: Metric kJ/kmol·K @ 15°C, 101.325 kPa
Methane	CH_4	C_1	16.04	668	343	4,604	191	8.45	35.35
Ethane	C_2H_6	C_2	30.07	708	550	4,880	305	12.27	51.30
Propane	C_3H_8	C_3	44.10	616	666	4,249	370	17.14	71.66
n-Butane	C_4H_{10}	nC_4	58.12	551	765	3,797	425	22.96	96.01
Isobutane	C_4H_{10}	iC_4	58.12	529	735	3,648	408	22.48	93.92
n-Pentane	C_5H_{12}	nC_5	72.15	489	845	3,369	470	27.99	117.03
Isopentane	C_5H_{12}	iC_5	72.15	490	829	3,381	460	27.63	115.44
n-Hexane	C_6H_{14}	nC_6	86.18	437	913	3,012	507	33.24	139.01
n-Heptane	C_7H_{16}	nC_7	100.20	397	972	2,736	540	38.50	160.92
n-Octane	C_8H_{18}	nC_8	114.23	361	1,024	2,486	569	43.76	182.88
n-Nonane	C_9H_{20}	nC_9	128.26	332	1,070	2,288	595	49.02	204.96
n-Decane	$C_{10}H_{22}$	nC_{10}	142.29	304	1,111	2,099	617	54.30	226.95
Ethylene	C_2H_4	C_2^-	28.05	731	508	5,041	282	10.16	42.47
Propylene (Propene)	C_3H_6	C_3^-	42.08	667	657	4,600	365	14.90	62.28
Butylene (*1*-Butene)	C_4H_8	nC_4^-	56.11	584	755	4,023	420	19.91	83.21
Cis-2-Butene	C_4H_8		56.11	612	784	4,220	436	18.34	76.65
Trans-2-Butene	C_4H_8		56.11	587	771	4,047	429	20.50	85.74
Isobutylene (Isobutene)	C_4H_8	iC_4^-	56.11	580	752	3,999	418	20.77	86.80
Pentylene (*1*-Pentene)	C_5H_{10}	C_5^-	70.14	512	837	3,529	465	25.50	106.54
1, 2 Butadiene	C_4H_6		54.09	653	800	4,502	444	18.70	78.21
1, 3 Butadiene	C_4H_6		54.09	628	765	4,330	425	18.46	77.13

*Values derived by permission from Gas Processors Suppliers Association *Engineering Data Book,* Ninth Edition, 1972, 5th Revision, 1981, and *SI Engineering Data Book,* 1980.

Table C-1 (continued)

Compound	Formula	Hydrocarbon Reference Symbol	Molecular Weight	Critical Constants: English Pressure, psia	Critical Constants: English Temperature, °R	Critical Constants: Metric Pressure, kPa	Critical Constants: Metric Temperature, K	Specific Heat, c_p: English Btu/lbm-mol-°R @ 60°F, 14.696 psia	Specific Heat, c_p: Metric kJ/kmol·K @ 15°C, 101.325 kPa
Isoprene	C_5H_8		68.12	558	872	3,850	484	24.32	101.64
Acetylene	C_2H_2	$C_2^=$	26.04	890	555	6,139	308	10.33	43.20
Benzene	C_6H_6		78.11	710	1,012	4,898	562	18.92	79.20
Toluene	C_7H_8		92.14	596	1,065	4,106	592	23.94	99.97
o-Xylene	C_8H_{10}		106.17	542	1,135	3,734	630	30.94	129.32
m-Xylene	C_8H_{10}		106.17	513	1,111	3,536	617	29.54	123.48
p-Xylene	C_8H_{10}		106.17	509	1,109	3,511	616	29.40	122.84
Styrene	C_8H_8		104.15	580	1,166	3,999	648	28.18	118.00
Methyl alcohol	CH_4O		32.04	1,174	923	8,096	513	10.35	43.32
Ethyl alcohol	C_2H_6O		46.07	926	925	6,383	514	15.29	63.99
Carbon monoxide	CO		28.01	508	239	3,499	133	6.96	29.13
Carbon dioxide	CO_2		44.01	1,071	547	7,382	304	8.76	36.66
Hydrogen sulfide	H_2S		34.08	1,036	672	9,005	374	8.11	33.94
Sulfur dioxide	SO_2		64.06	1,145	775	7,894	431	9.28	38.83
Ammonia	NH_3		17.03	1,636	730	11,280	406	8.46	35.41
Air	$N_2 + O_2$		28.96	547	238	3,771	132	6.95	29.10
Hydrogen	H_2		2.02	188	60	1,297	33	6.87	28.76
Oxygen	O_2		32.00	737	278	5,081	155	7.00	29.33
Nitrogen	N_2		28.01	493	227	3,399	126	6.96	29.13
Chlorine	Cl_2		70.91	1,118	751	7,711	417	8.06	33.75
Water	H_2O		18.02	3,208	1,165	22,118	647	8.01	33.55
Helium	He		4.00	33	9	228	5	4.96	20.77
Hydrogen chloride	HCl		36.46	1,205	584	8,309	325	6.96	29.14

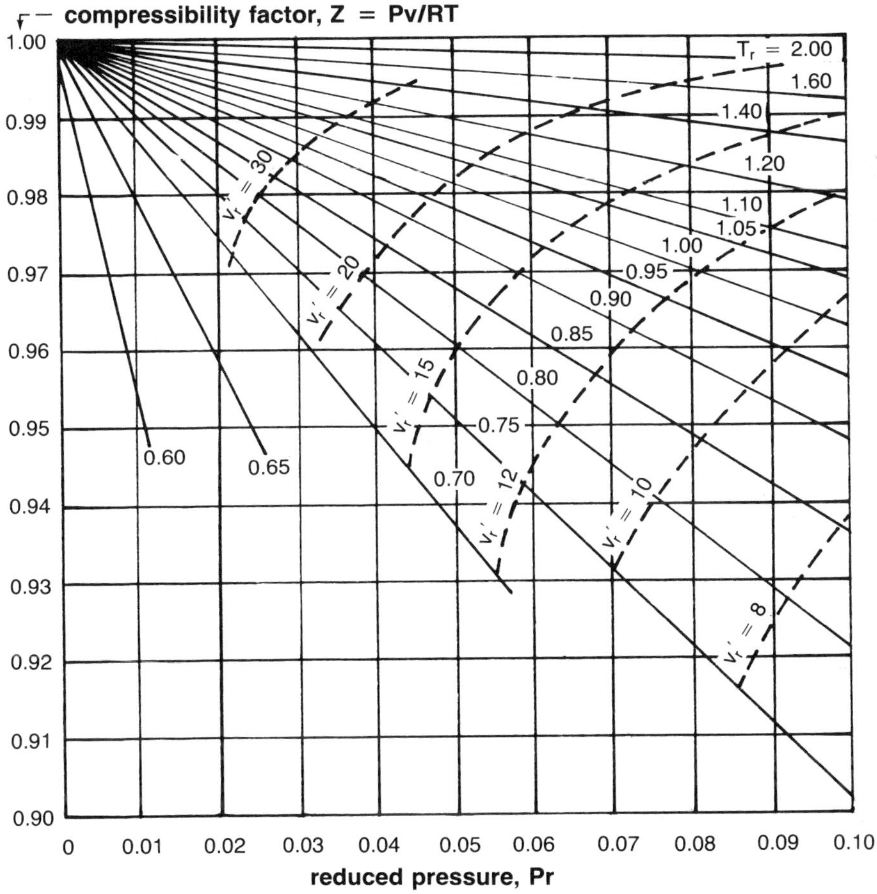

Figure C-1. Compressibility chart for very low values of reduced pressure. Reprinted by permission of *Chemical Engineering*, McGraw-Hill Publications Company, July, 1954.

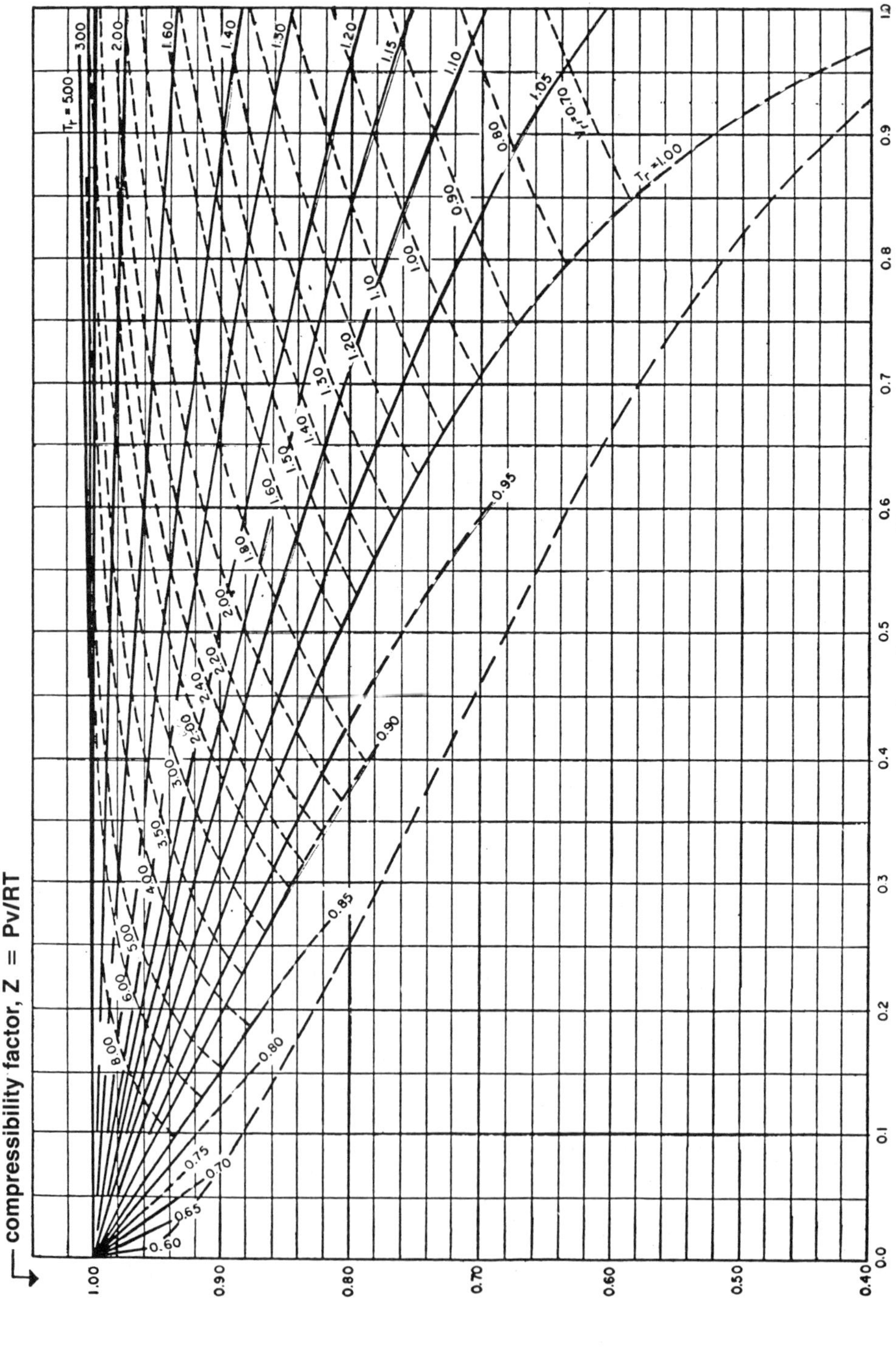

Figure C-2. Compressibility chart for low values of reduced pressure. Reprinted by permission of *Chemical Engineering*, McGraw-Hill Publications Co., July 1954.

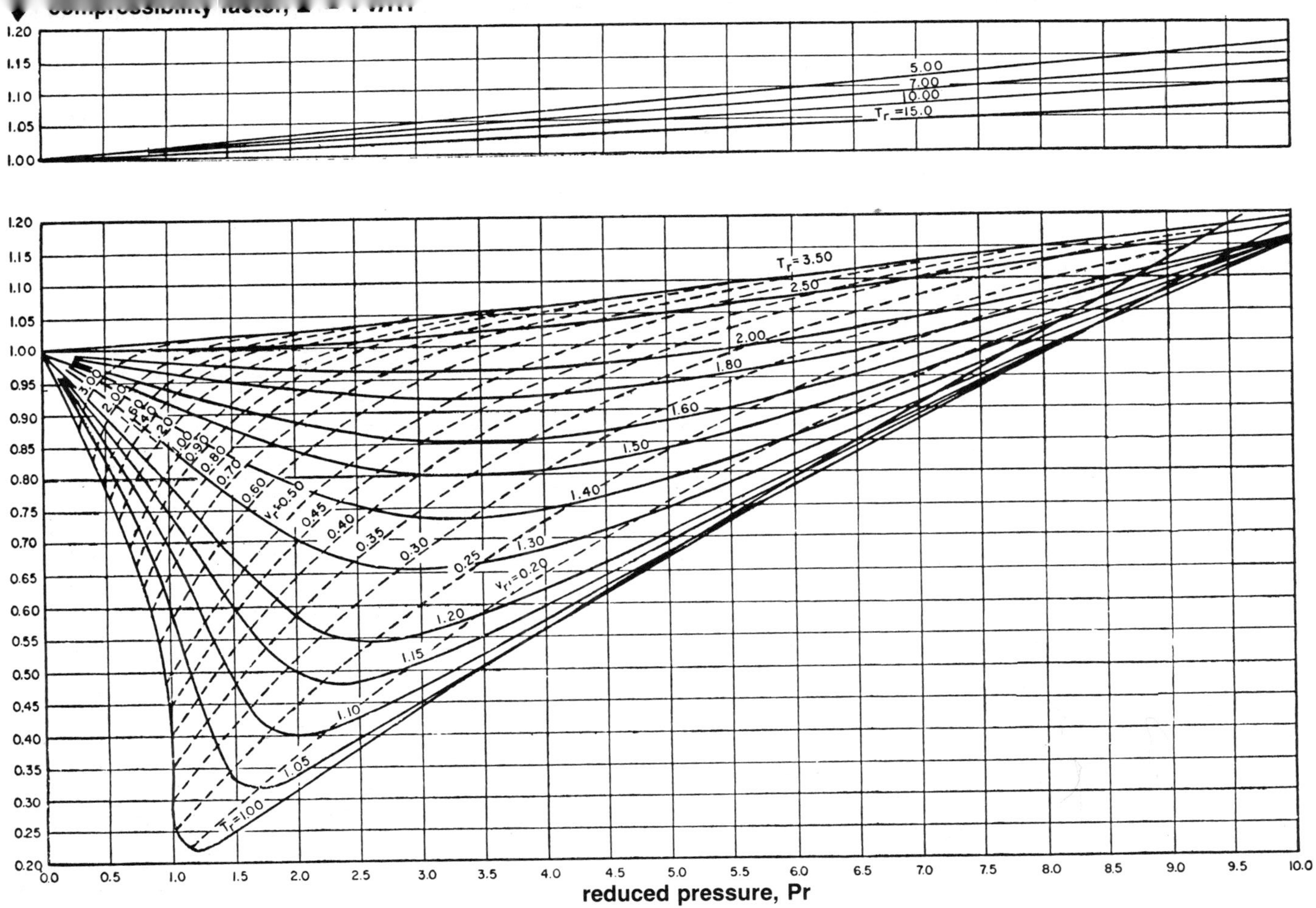

Figure C-3. Compressibility chart for low to high values of reduced pressure. Reprinted by permission of *Chemical Engineering*, McGraw-Hill Publications Co., July 1954.

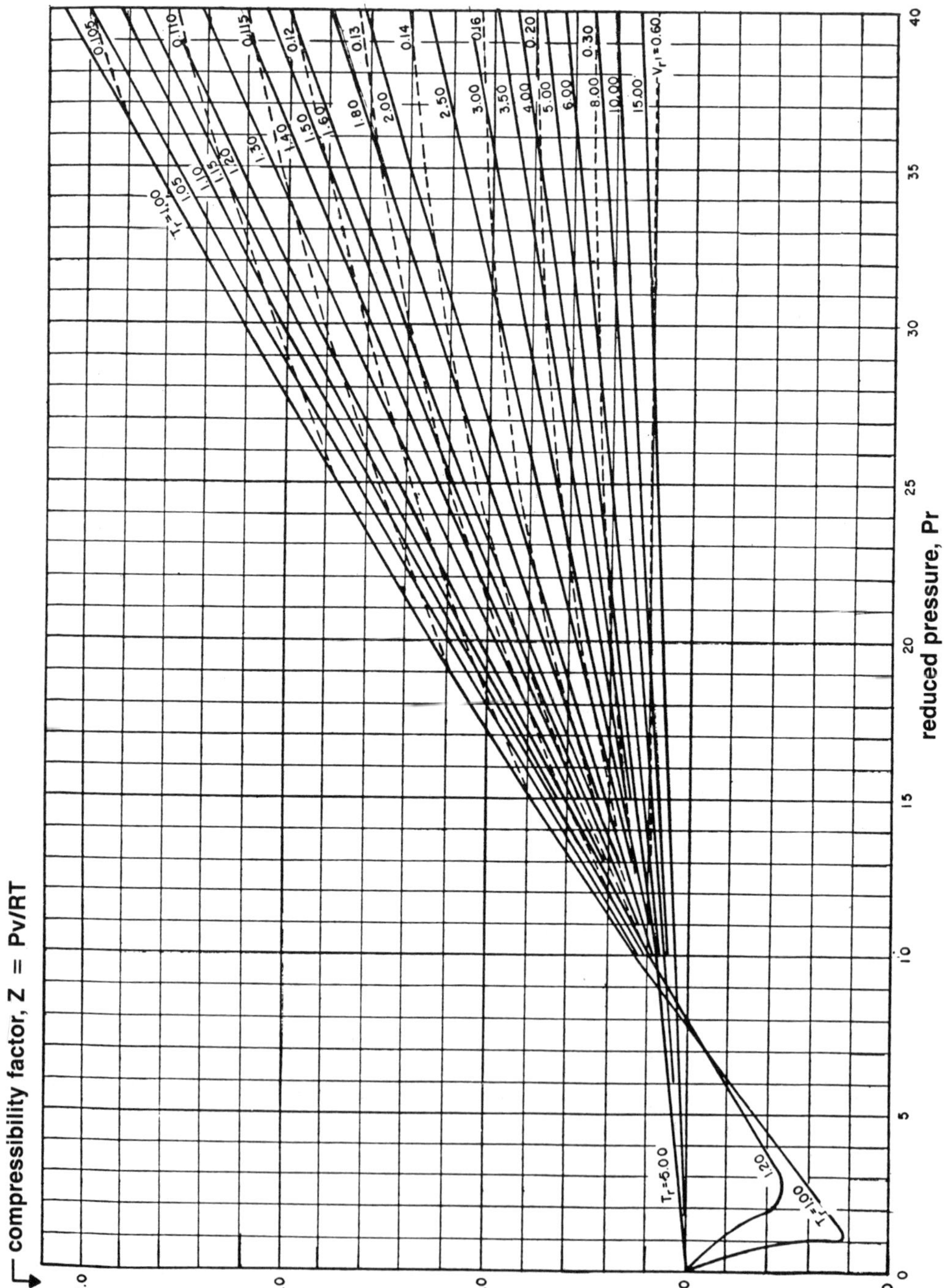

Figure C-4. Compressibility chart for low to very high values of reduced pressure. Reprinted by permission of Chemical Engineering, McGraw-Hill Publications Co., July 1954.

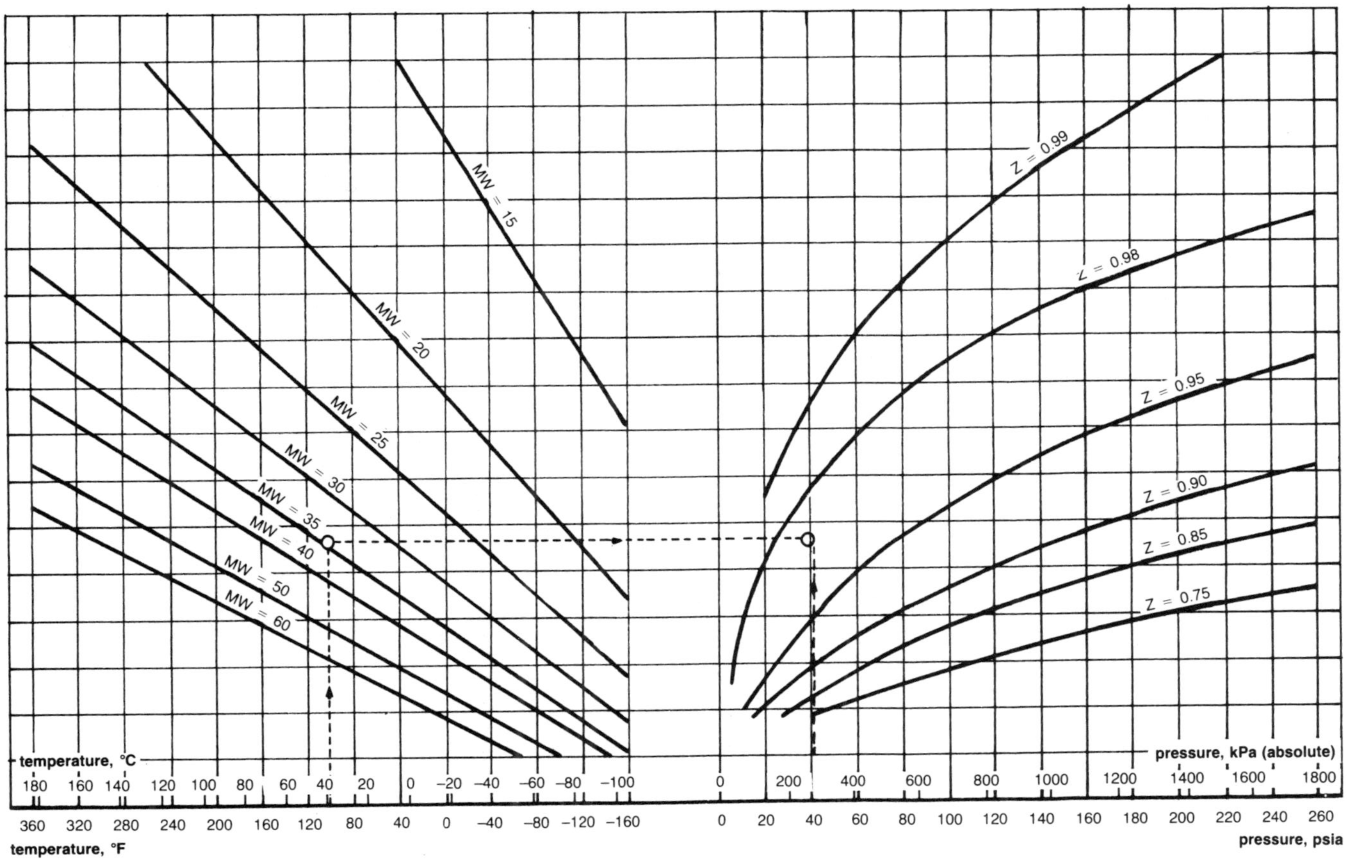

Figure C-5. Quick estimate for compressibility of hydrocarbon mixtures. Redrawn by permission of Compressed Air and Gas Institute, Vol. 13, No. 2; and *Chemical Engineering,* McGraw-Hill Publications Company, Feb. 8, 1982.

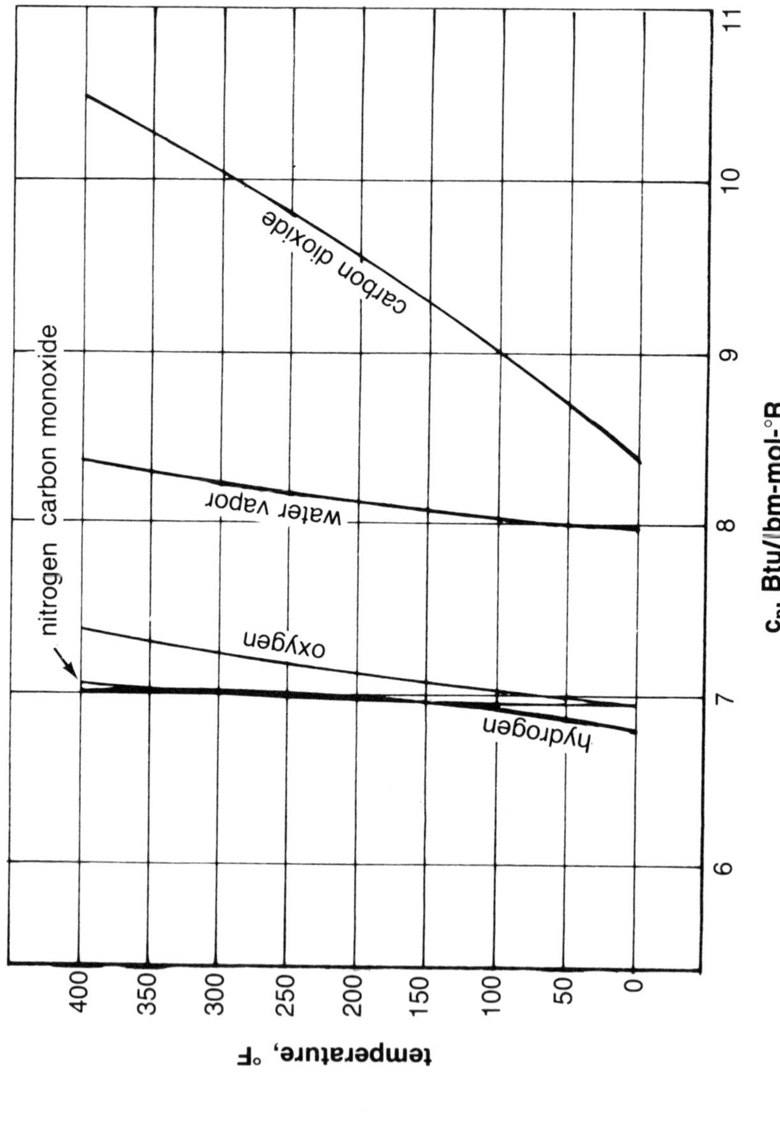

Figure C-6. Molal specific heat as a function of temperature for several nonhydrocarbon gases—English system. Reference data developed and published by the Thermodynamics Research Center of Texas A&M University.

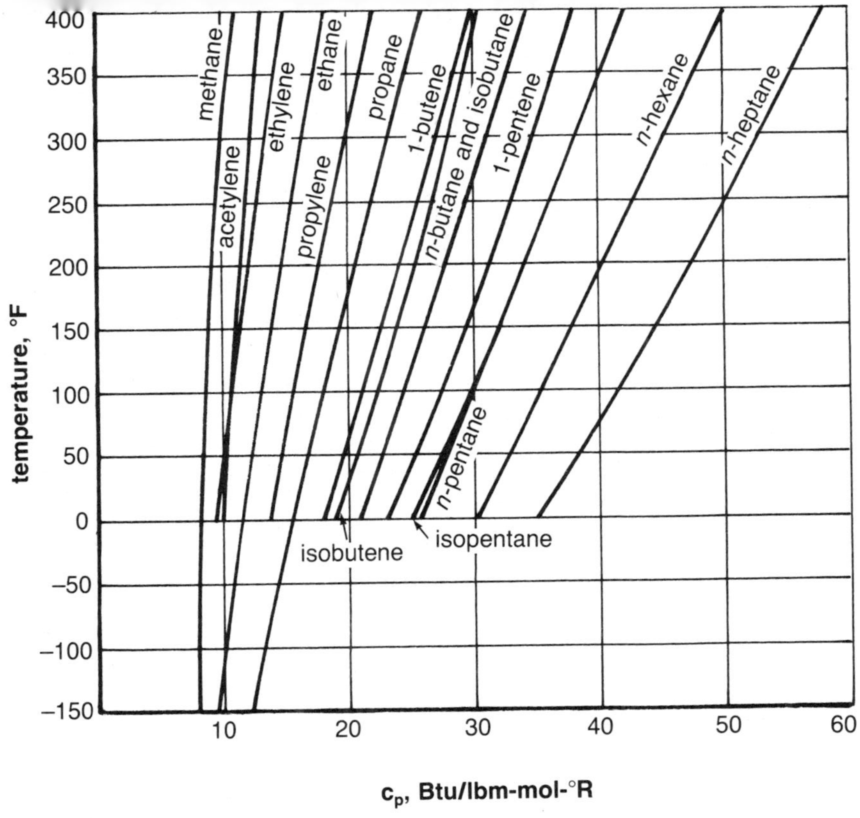

Figure C-7. Molal specific heat as a function of temperature for several hydrocarbon gases—English system. Reference data developed and published by the Thermodynamics Research Center of Texas A&M University.

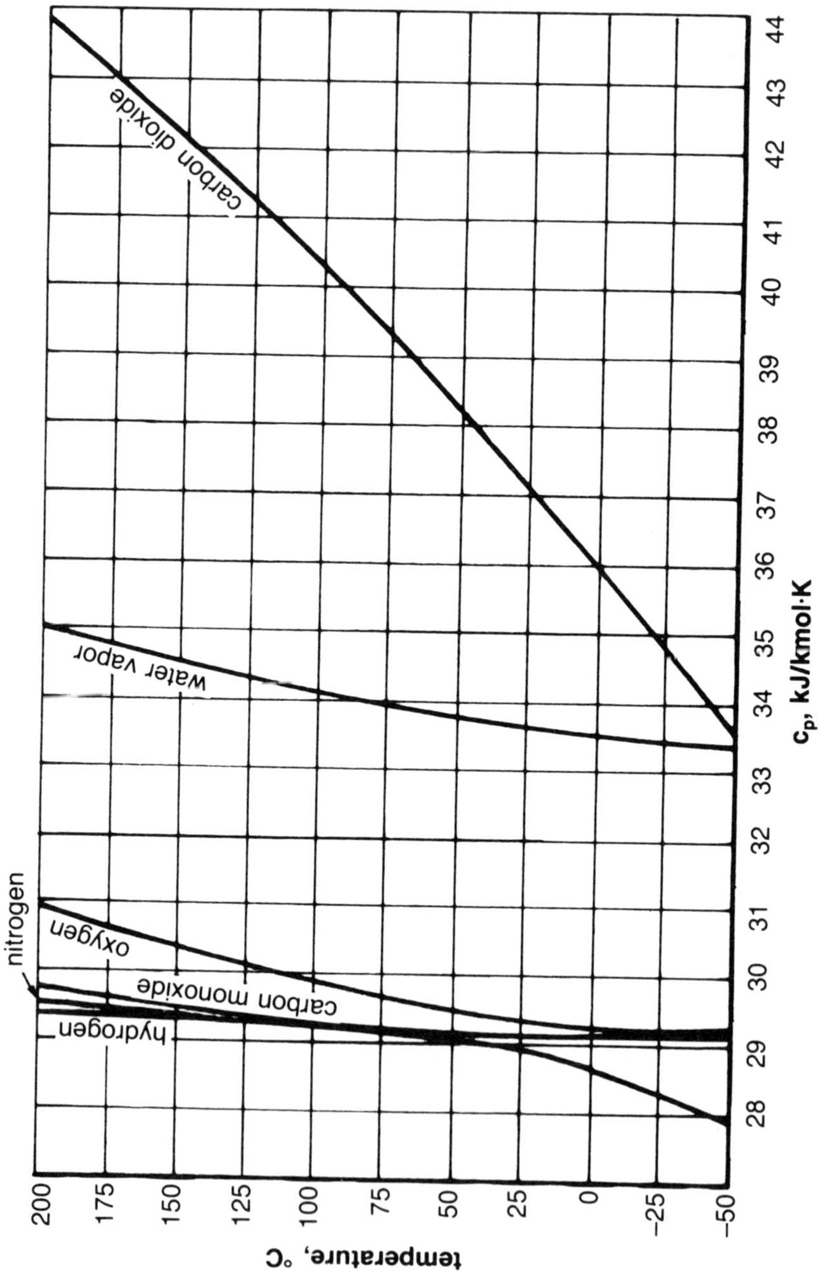

Figure C-8. Molal specific heat as a function of temperature for several nonhydrocarbon gases—metric system. Reference data developed and published by the Thermodynamics Research Center of Texas A&M University.

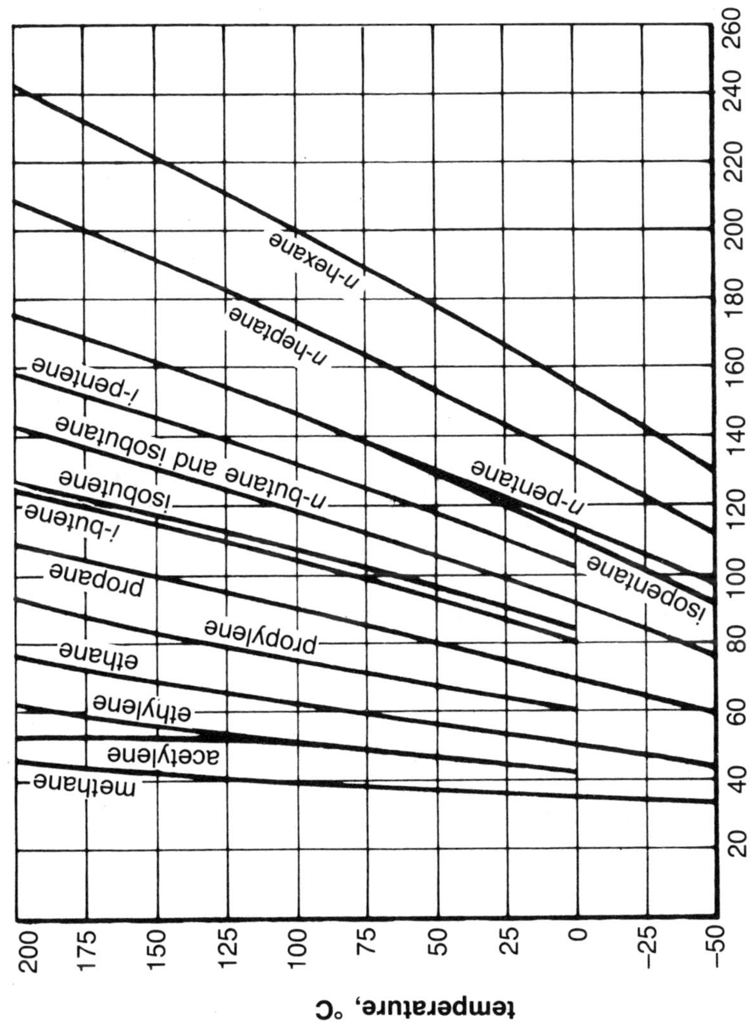

Figure C-9. Molal specific heat as function of temperature for several hydrocarbon gases—metric system. Reference data developed and published by the Thermodynamics Research Center of Texas A&M University.

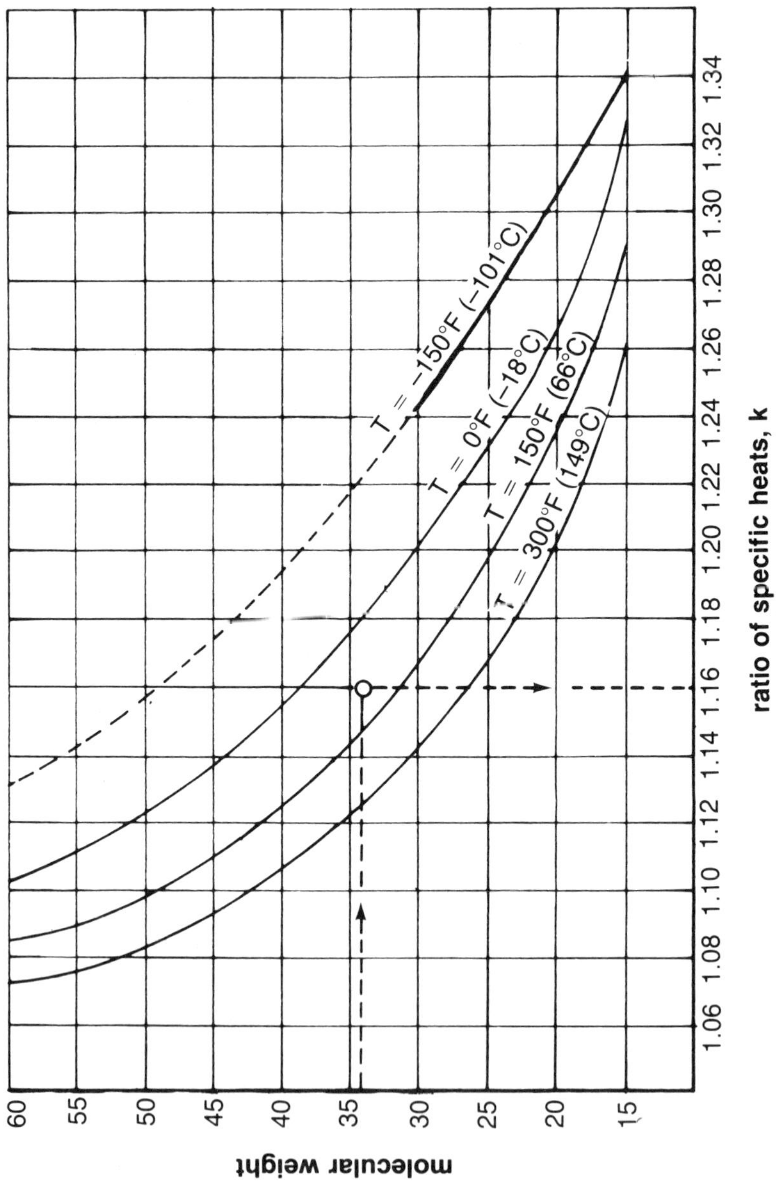

Figure C-10. Quick estimate for ratio of specific heats for hydrocarbon mixtures. Redrawn by permission of Compressed Air and Gas Institute, Vol. 13, No. 2; and *Chemical Engineering*, McGraw-Hill Publications Company, Feb. 8, 1982.

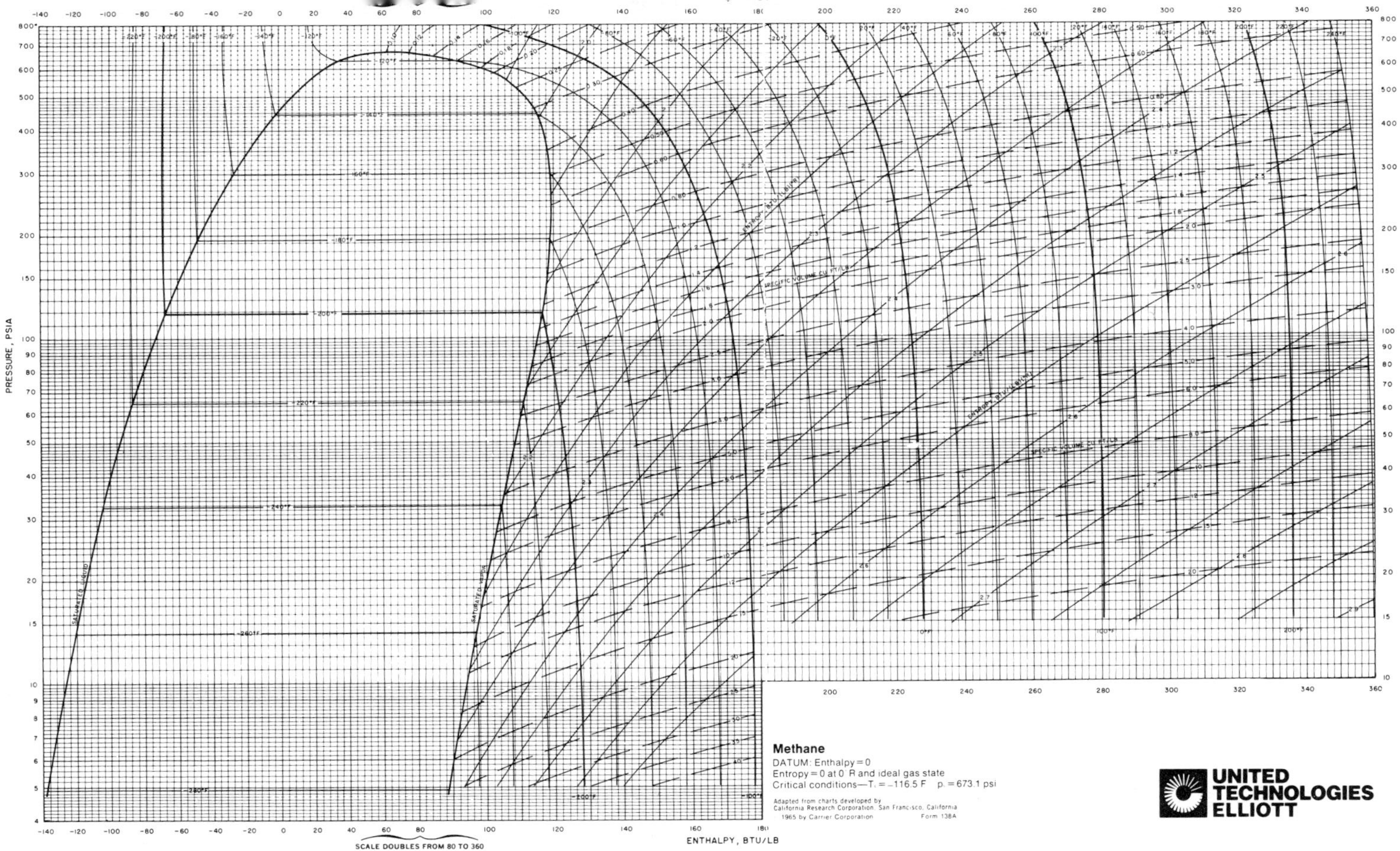

Figure C-11. Methane Mollier chart—English system. Reprinted by permission of Elliott Company, Jeannette, PA, Bulletin P-25S-Rev, 1981.

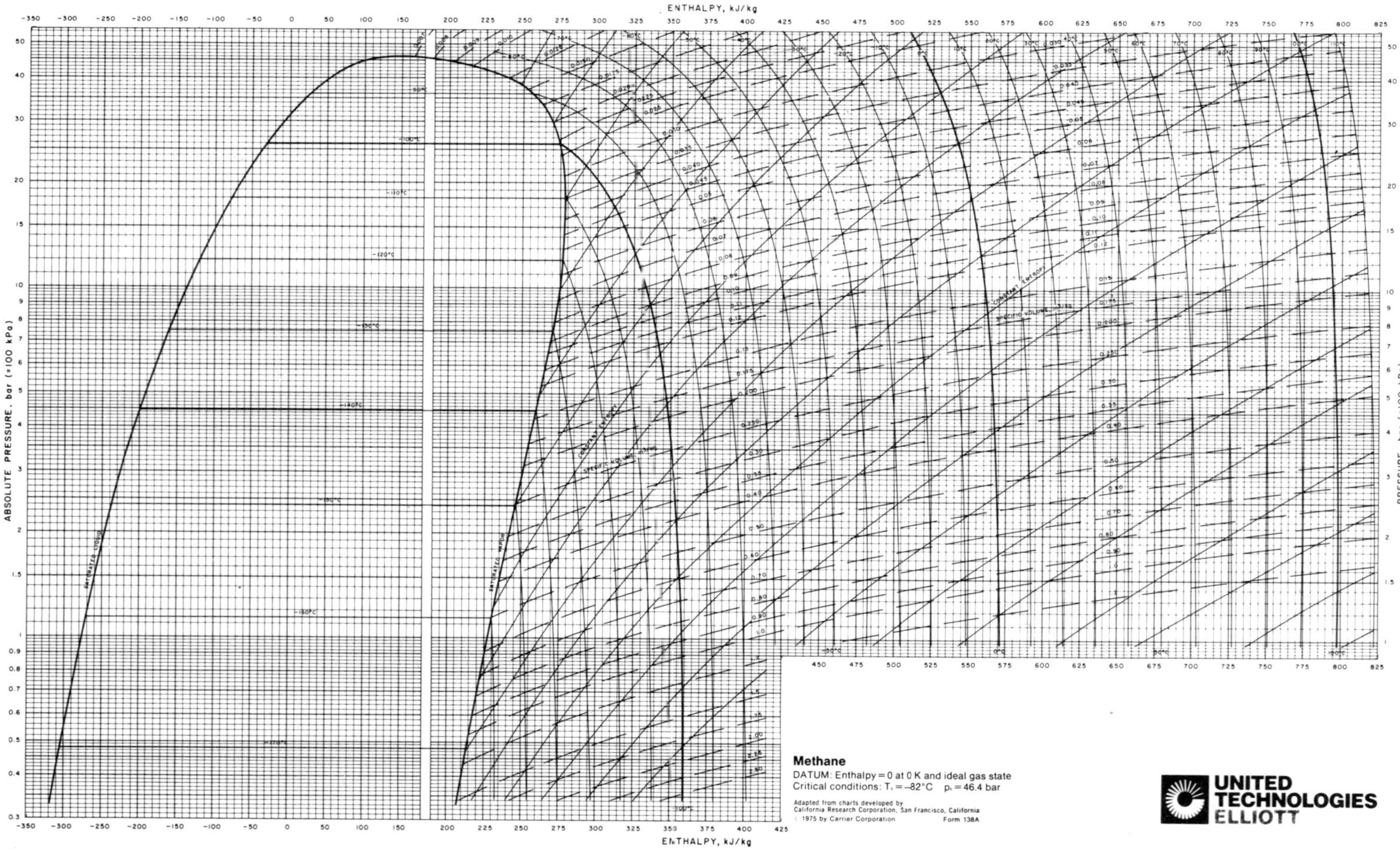

Figure C-12 Methane Mollier chart—metric system. Reprinted by permission of Elliott Company, Jeannette, PA, Bulletin P- 25S-Rev, 1981.

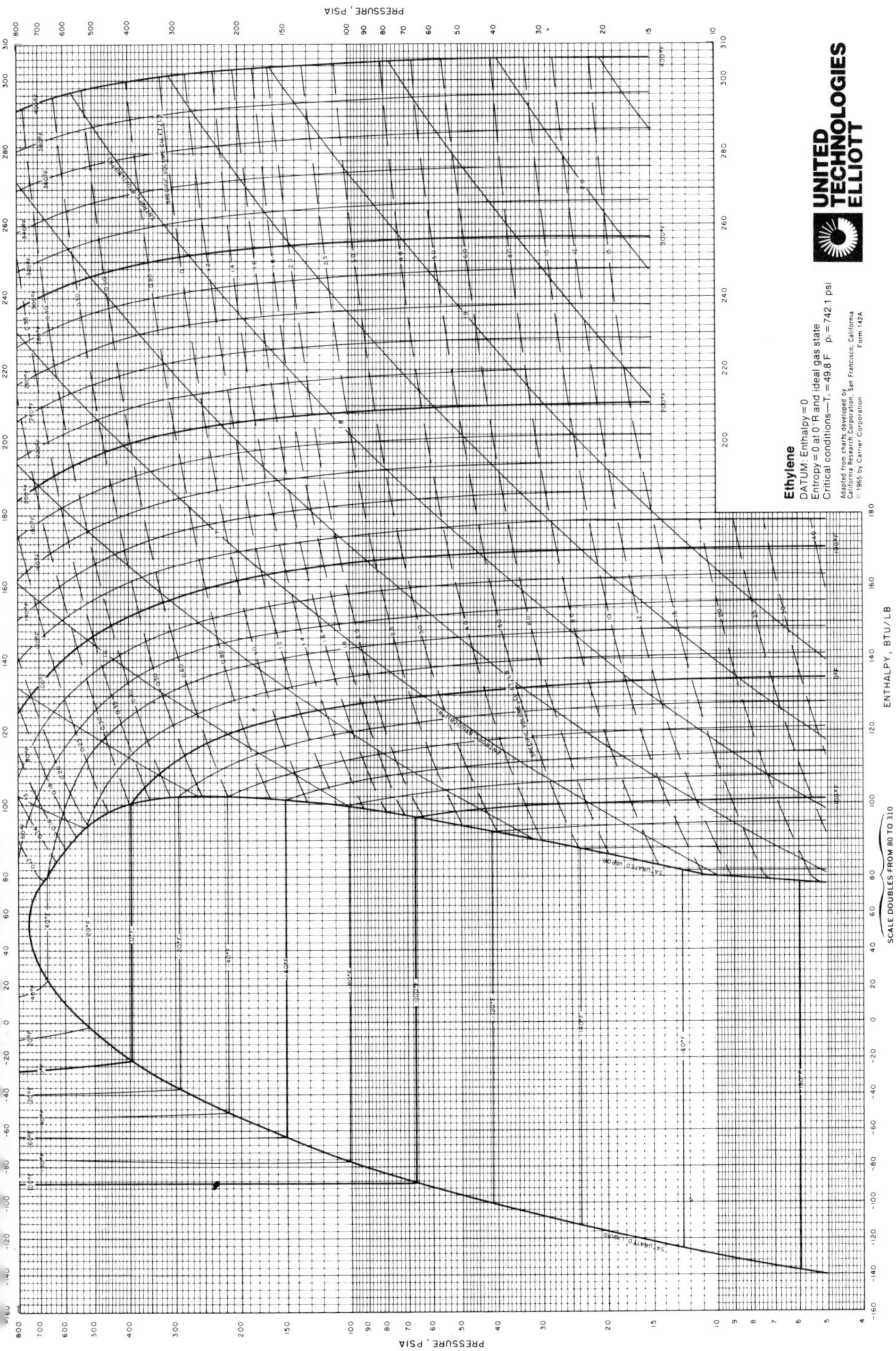

Figure C-13. Ethylene Mollier chart—English system. Reprinted by permission of Elliott Company, Jeannette, PA, Bulletin P-25S-Rev, 1981.

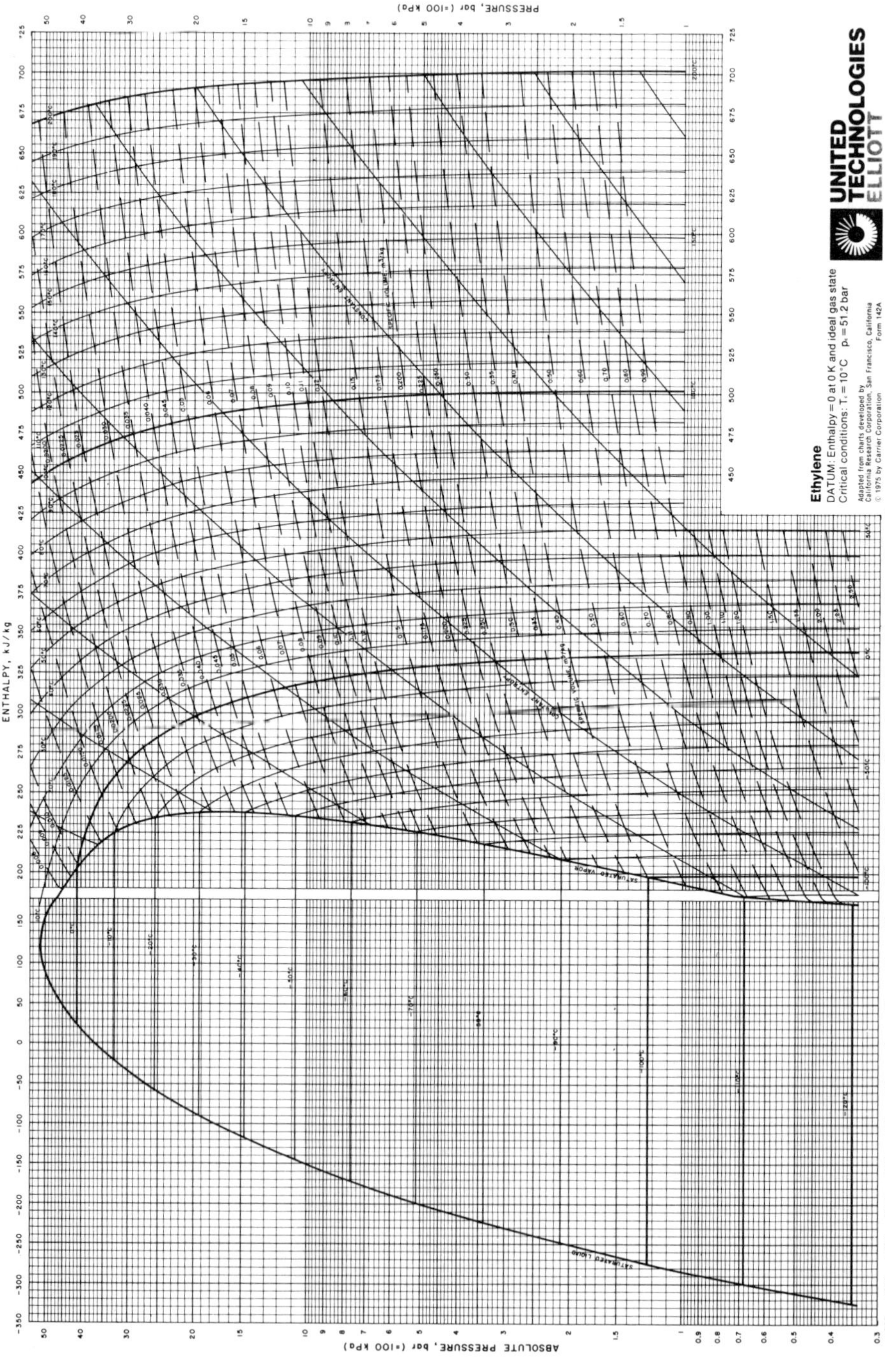
PRESSURE, bar (=100 kPa)
ENTHALPY, kJ/kg
ABSOLUTE PRESSURE, bar (=100 kPa)
SATURATED VAPOR
SATURATED LIQUID
CONSTANT ENTROPY
SPECIFIC VOLUME, m³/kg
Ethylene
DATUM: Enthalpy = 0 at 0 K and ideal gas state
Critical conditions: Tc = 10°C pc = 51.2 bar
Adapted from charts developed by
California Research Corporation, San Francisco, California
© 1975 by Carrier Corporation
Form 142A
UNITED TECHNOLOGIES ELLIOTT

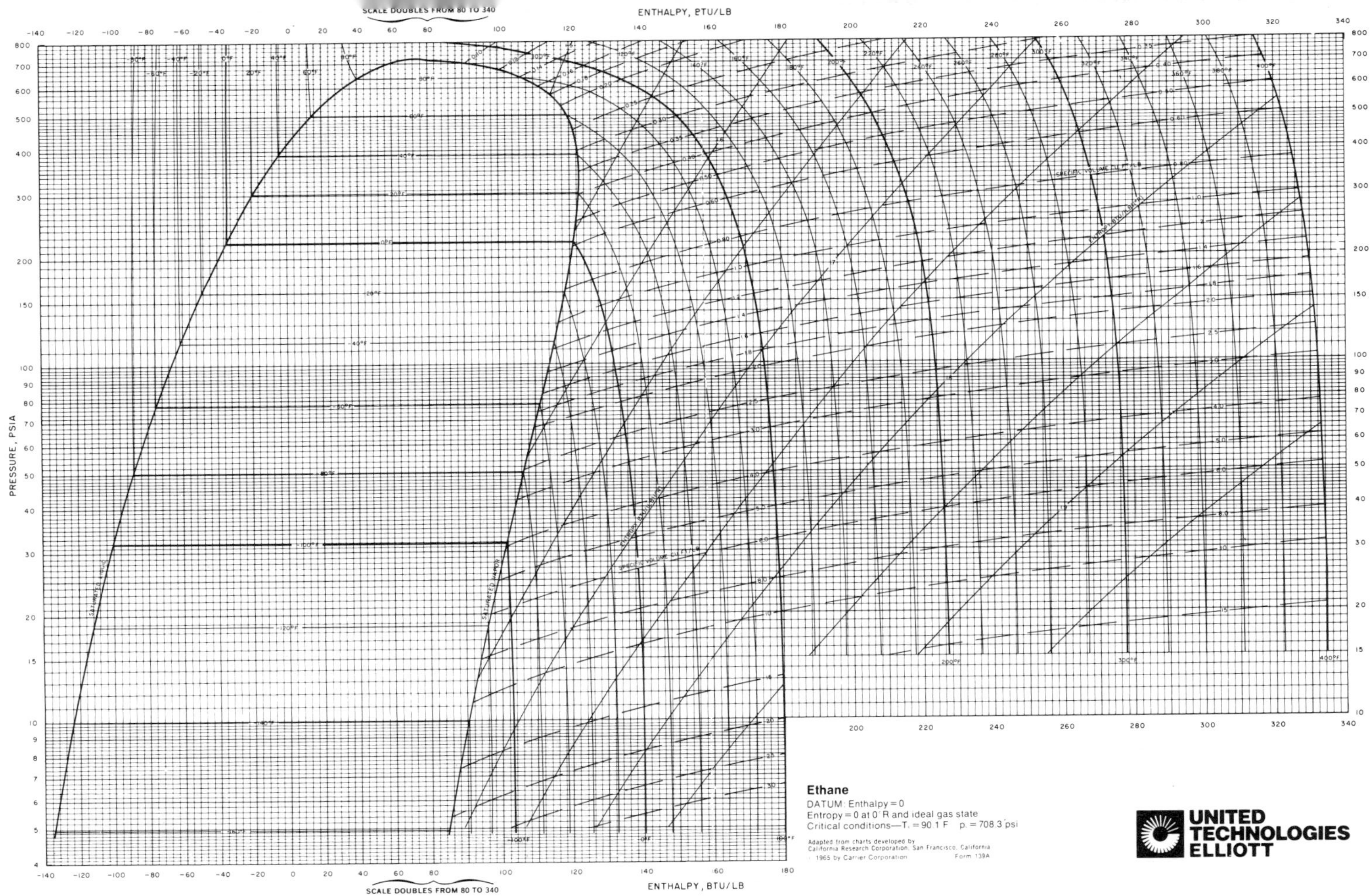

UNITED TECHNOLOGIES ELLIOTT

Figure C-15. Ethane Mollier chart—English system. Reprinted by permission of Elliott Companv, Jeannette, PA, Bulletin P- 25S-Rev, 1981.

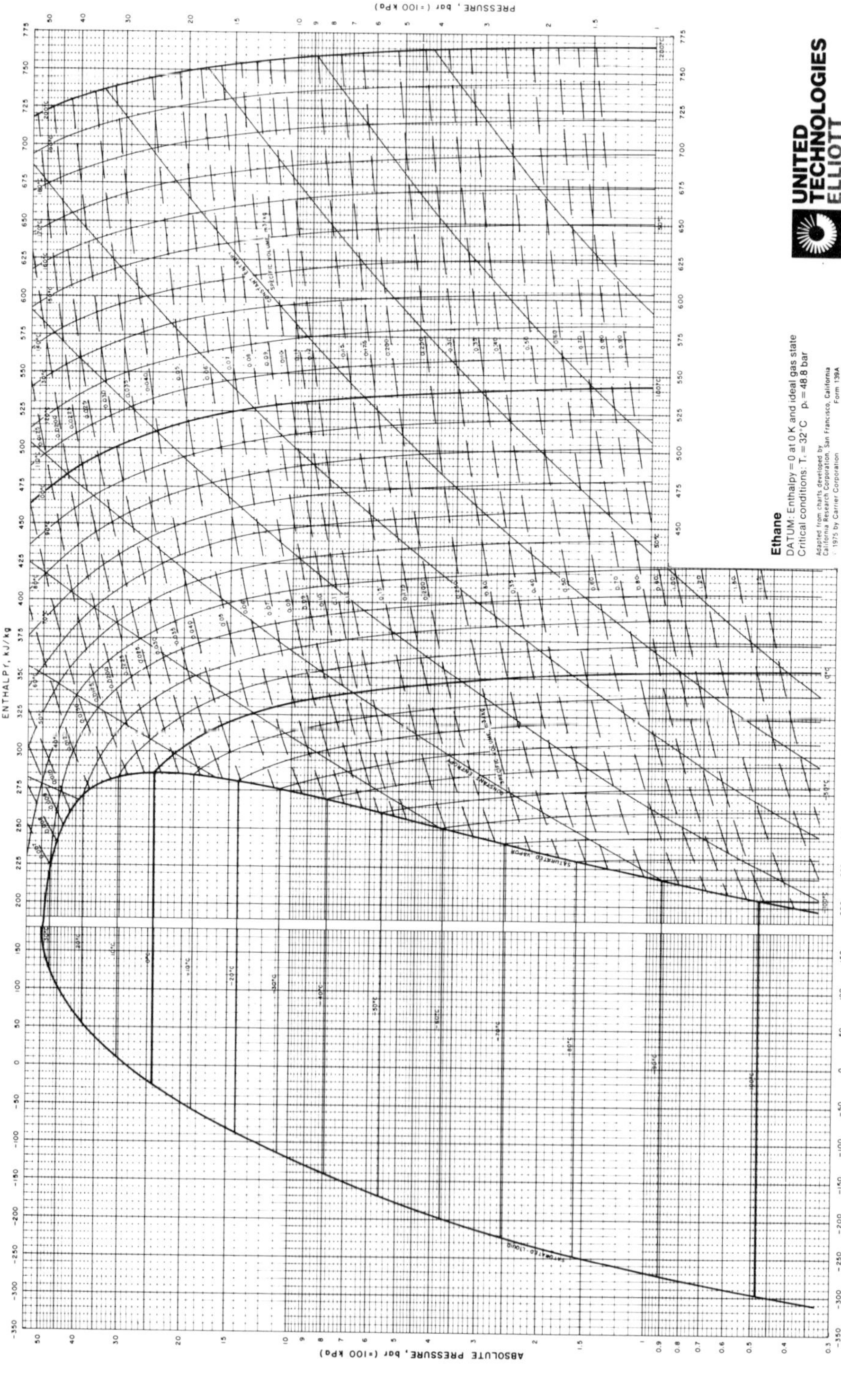
PRESSURE, bar (=100 kPa)
ENTHALPY, kJ/kg
ABSOLUTE PRESSURE, bar (=100 kPa)
SATURATED VAPOR
SATURATED LIQUID
CONSTANT ENTROPY
SPECIFIC VOLUME
Ethane
DATUM: Enthalpy = 0 at 0 K and ideal gas state
Critical conditions: T_c = 32°C p_c = 48.8 bar
Adapted from charts developed by California Research Corporation, San Francisco, California
© 1975 by Carrier Corporation Form 139A
UNITED TECHNOLOGIES ELLIOTT

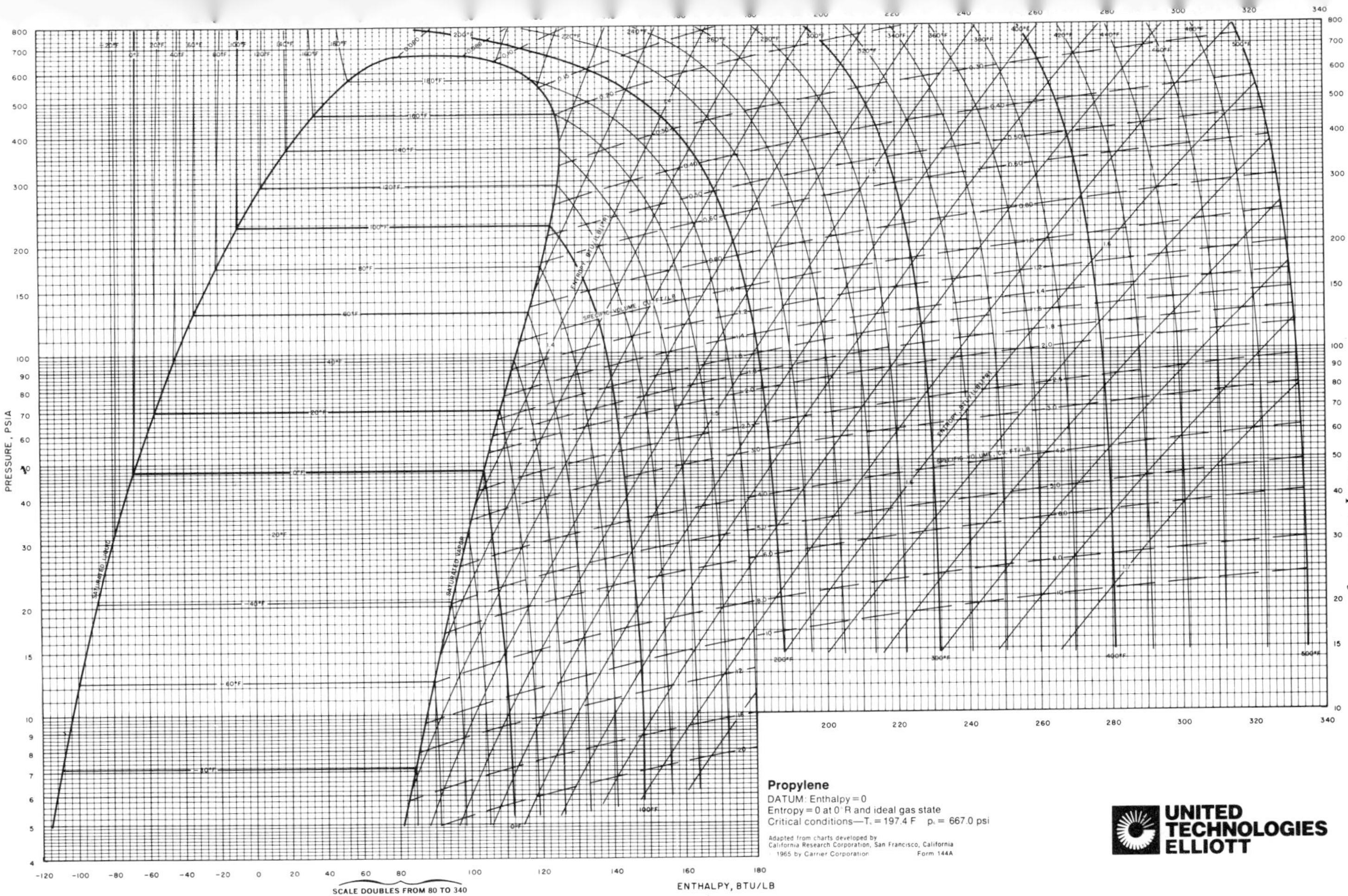

Figure C-17. Propylene Mollier chart—English system. Reprinted by permission of Elliott Company, Jeannette, PA, Bulletin P-25S-Rev, 1981.

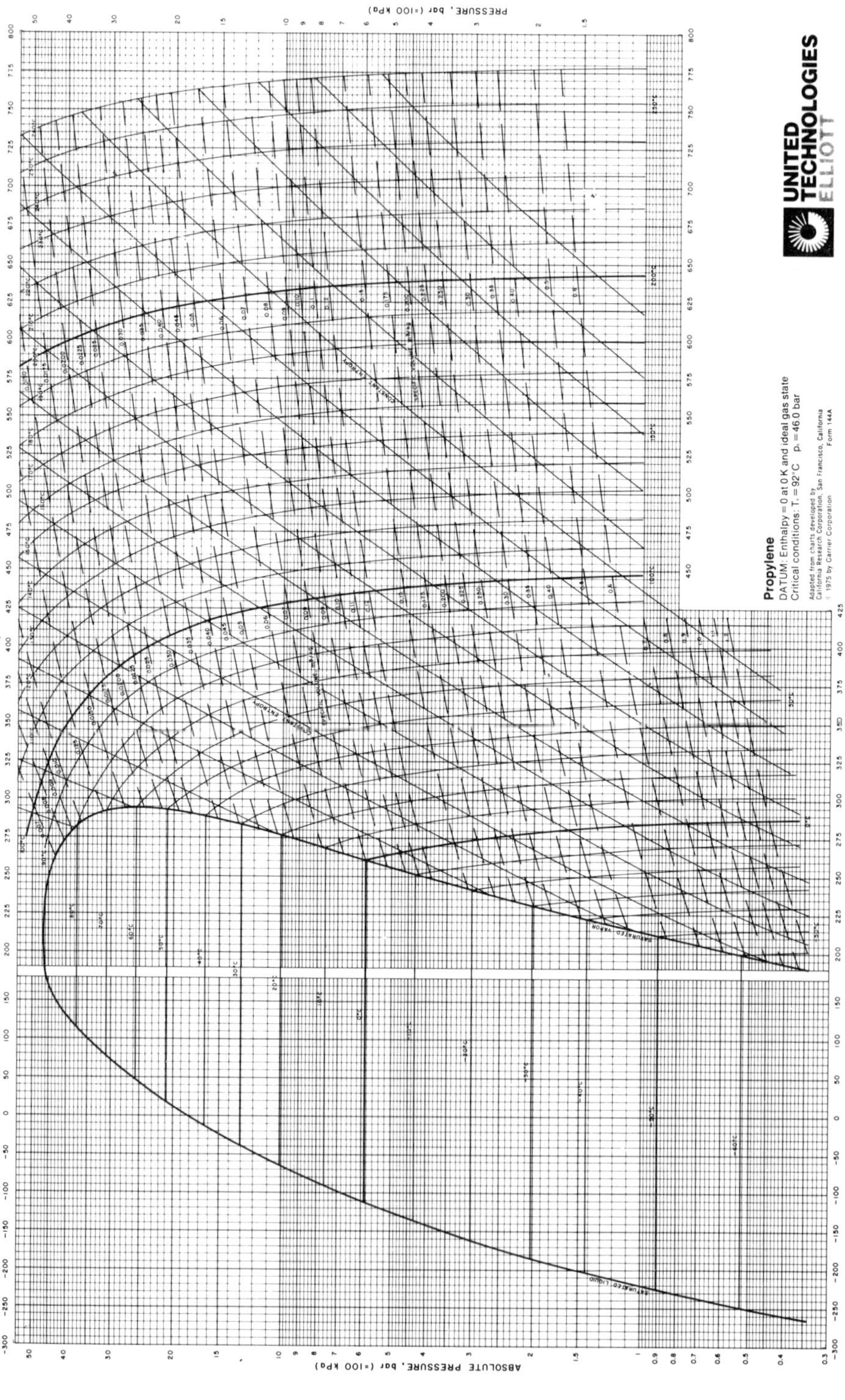

[illegible] Reprinted by permission of Elliott Company, Jeannette, PA, Bulletin

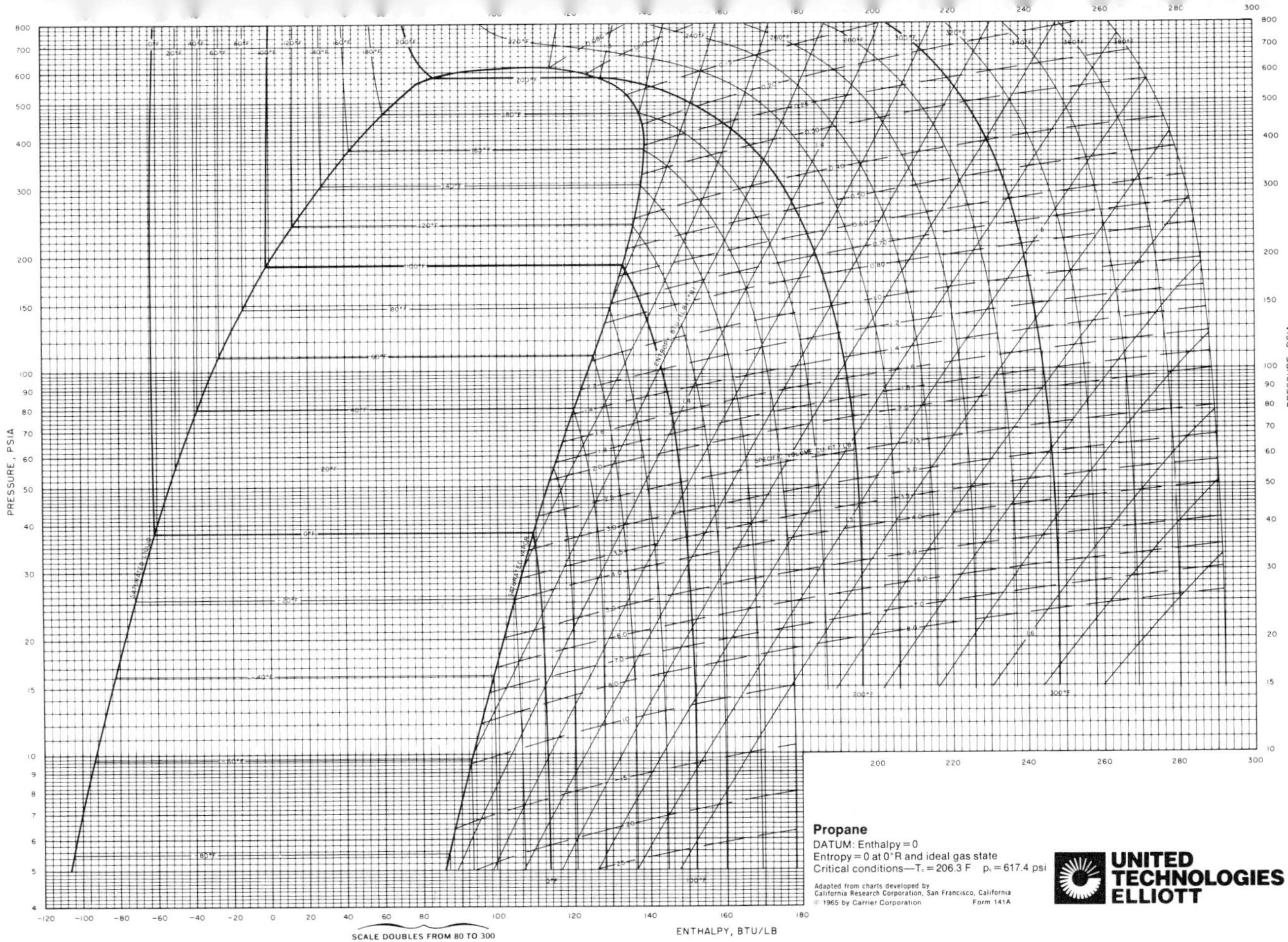

Figure C-19. Propane Mollier chart—English system. Reprinted by permission of Elliott Company, Jeannette, PA, Bulletin P-25S-Rev, 1981.

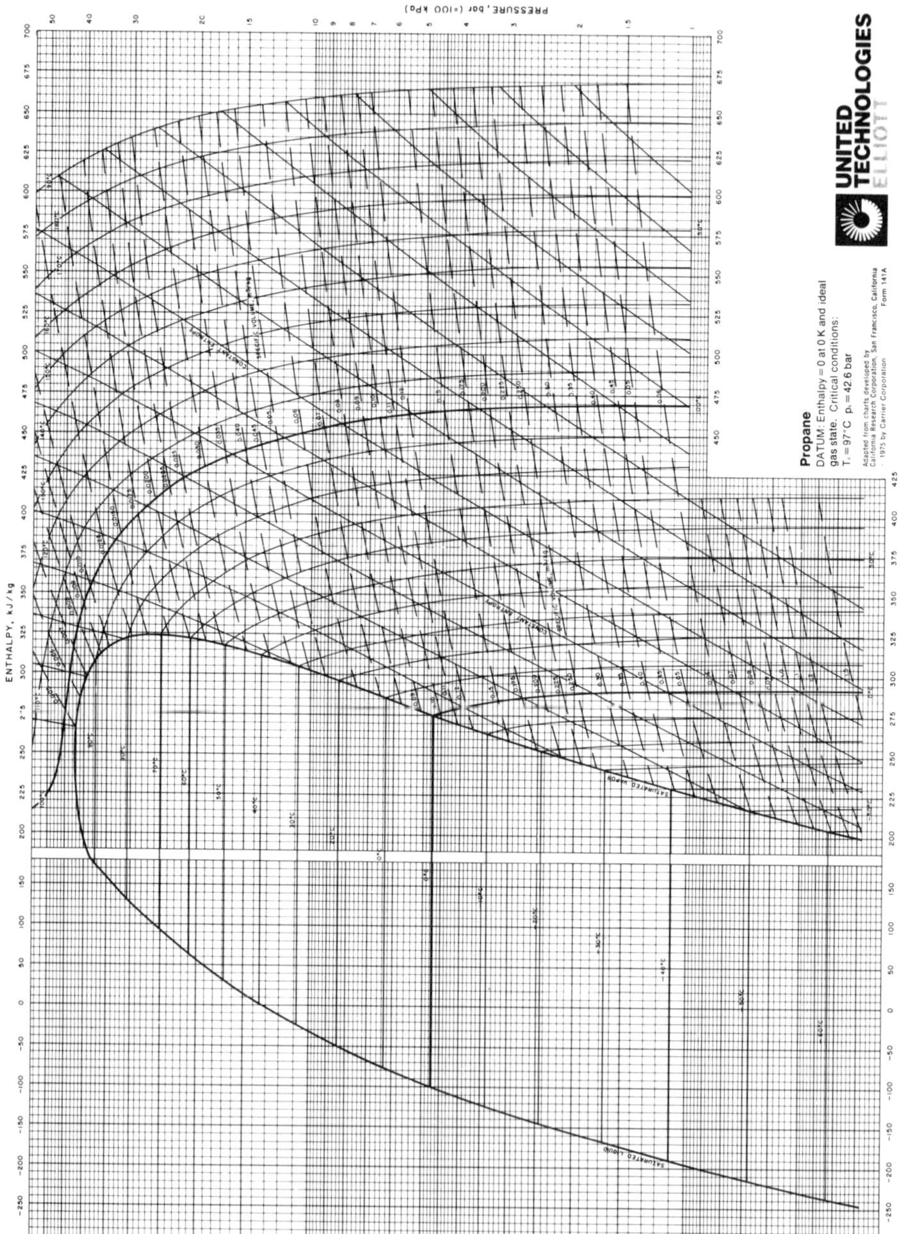

Figure C-20. Propane Mollier chart—metric system. Reprinted by permission of Elliott Company, Jeannette, PA, Bulletin P-25S-Rev, 1981.

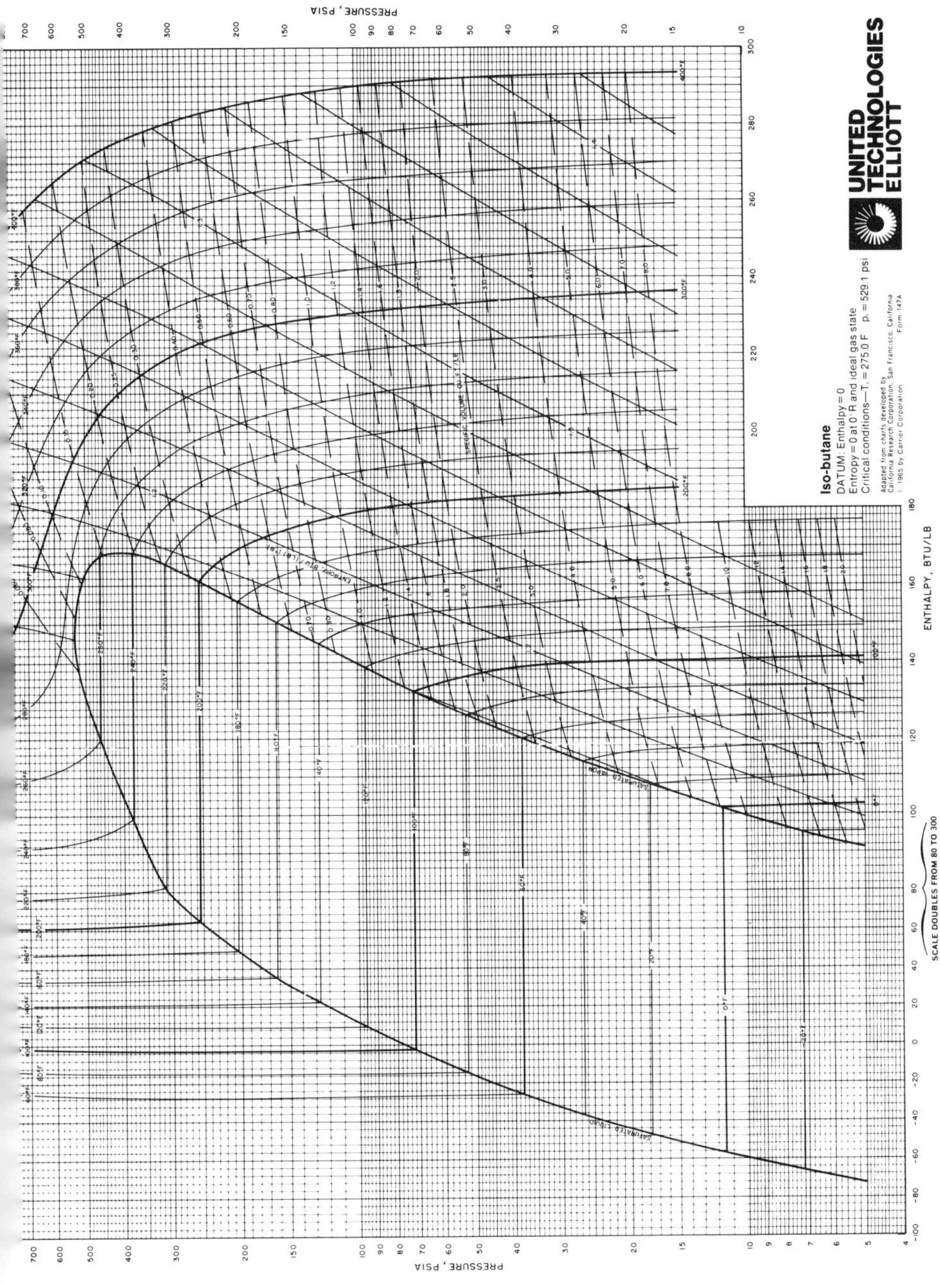

Figure C-21. Iso-butane Mollier chart—English system. Reprinted by permission of Elliott Company, Jeannette, PA, Bulletin P-25S-Rev, 1981.

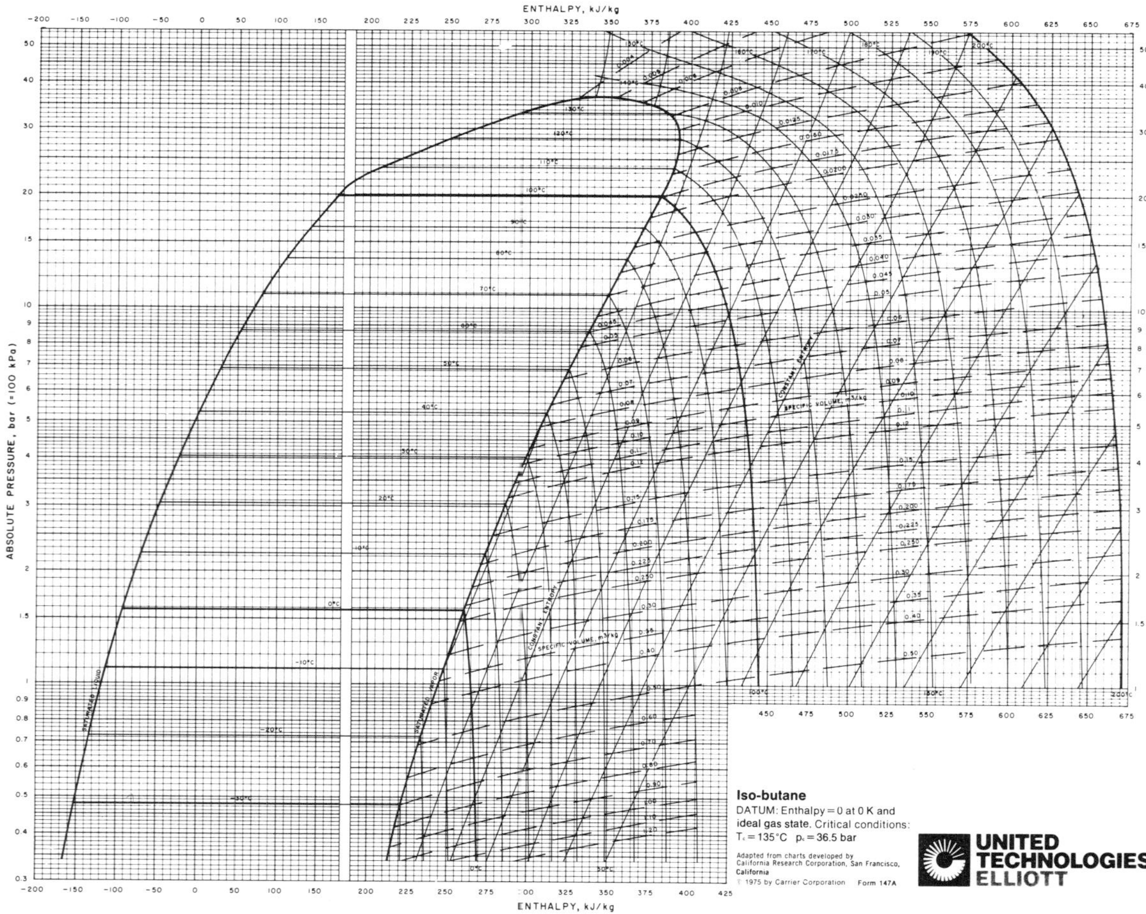

Figure C-22. Iso-butane Mollier chart—metric system. Reprinted by permission of Elliott Company, Jeannette, PA, Bulletin P-25S-Rev, 1981.

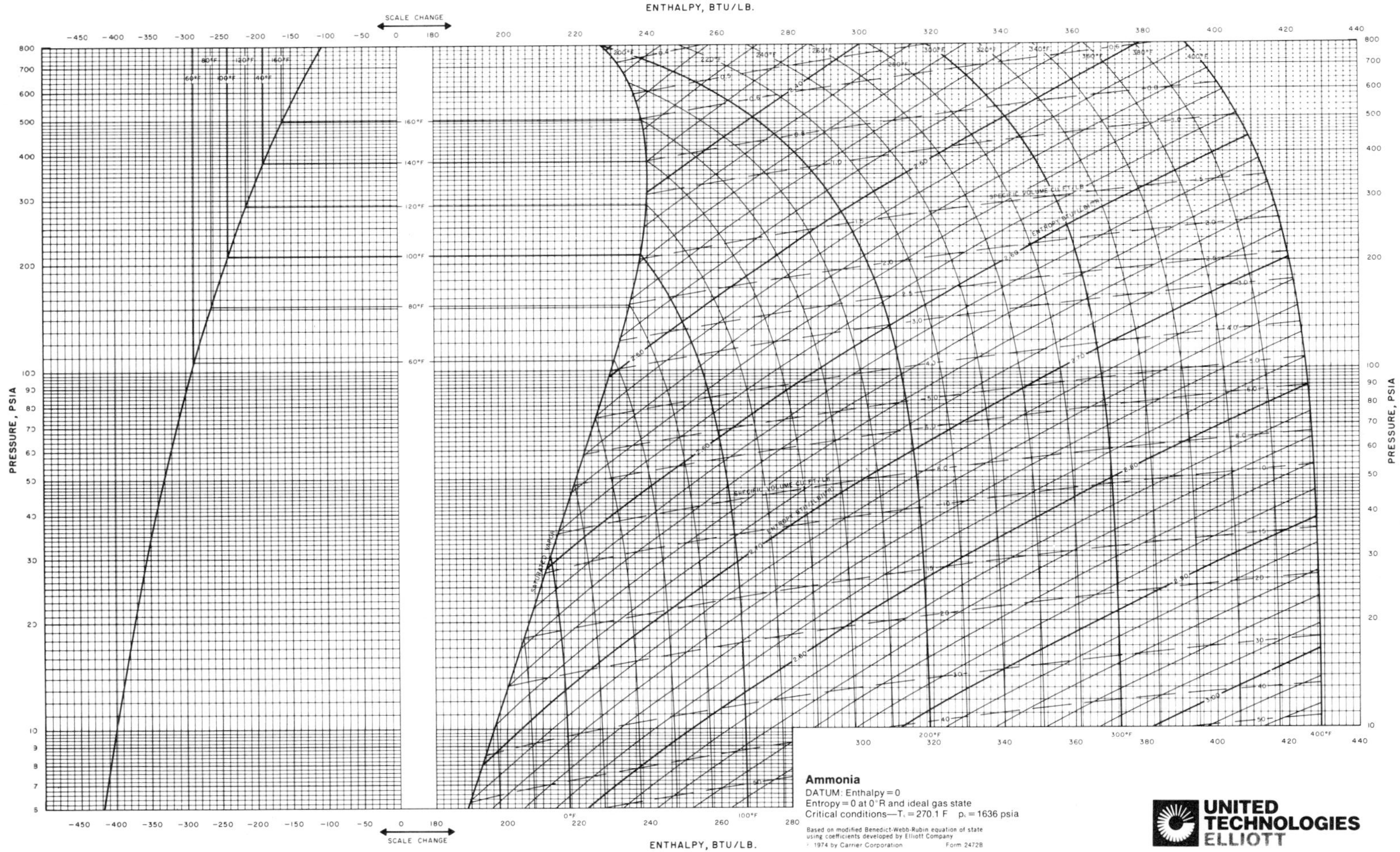

Figure C-23. Ammonia Mollier chart—English system. Reprinted by permission of Elliott Company, Jeannette, PA, Bulletin P-25S-Rev, 1981.

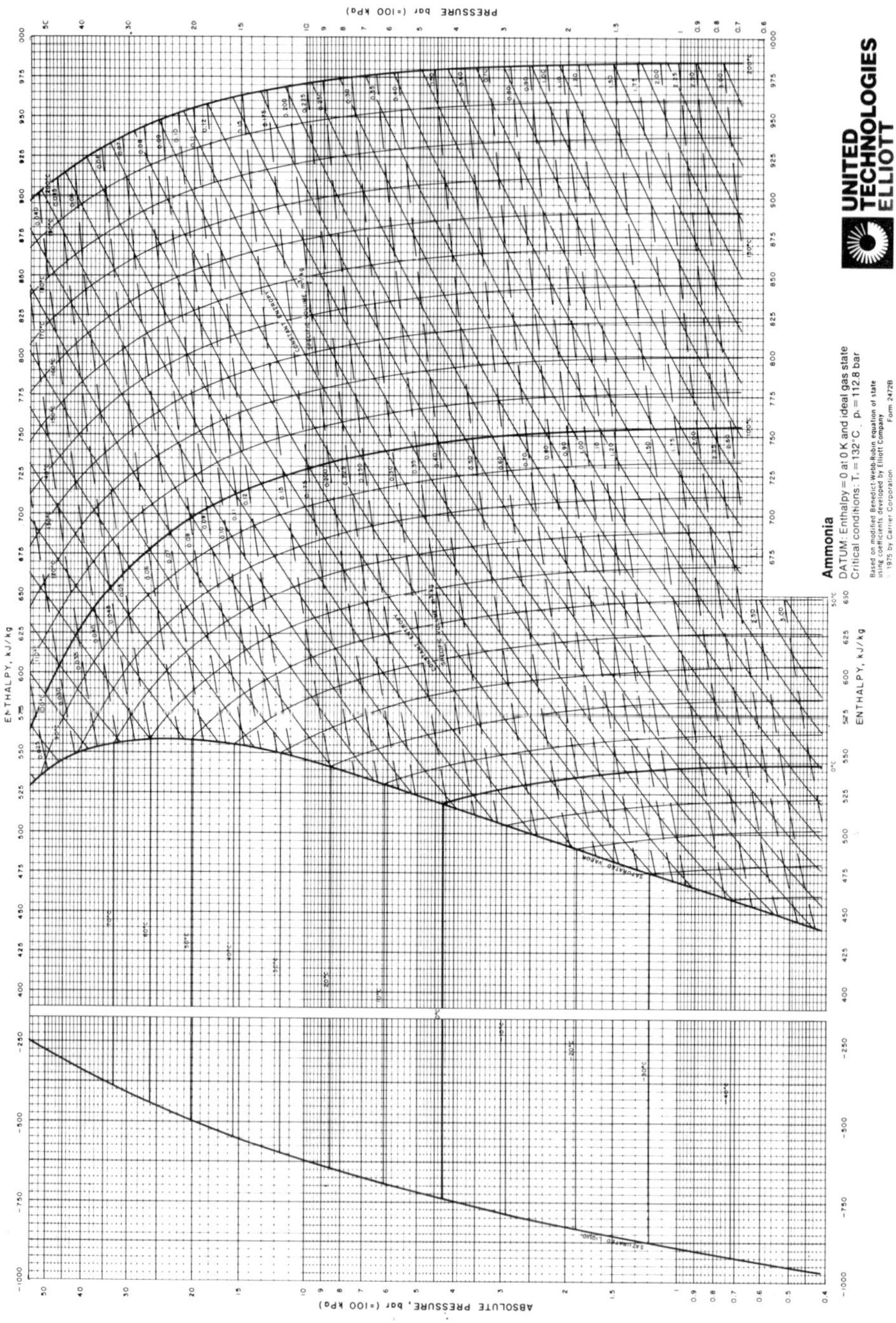

Figure C-24. Ammonia Mollier chart—metric system. Reprinted by permission of Elliott Company, Jeannette, PA, Bulletin P-25S-Rev, 1981.

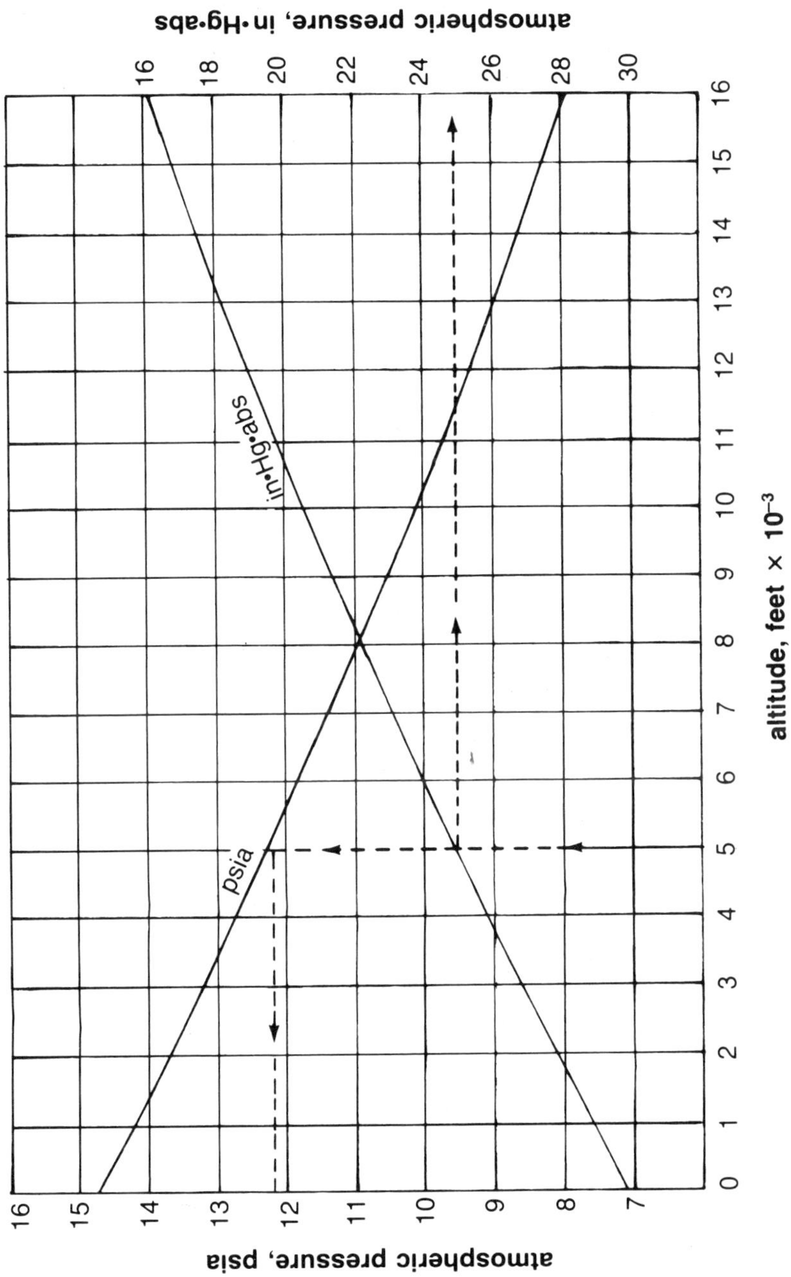

Figure C-25. Altitude to atmospheric pressure conversion chart—English system. Reference data published in *Aeronautical Vest-Pocket Handbook* by Pratt & Whitney Aircraft Group, 18th Edition, June, 1978.

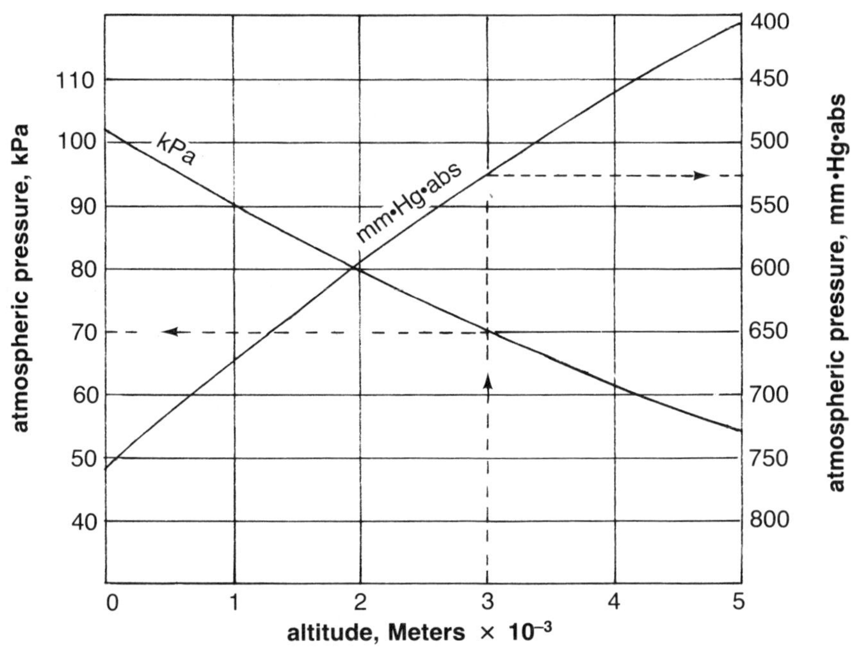

Figure C-26. Altitude to atmospheric pressure conversion chart—metric system. Reference data published in *Aeronautical Vest-Pocket Handbook* by Pratt & Whitney Aircraft Group, 18th Edition, June, 1978.

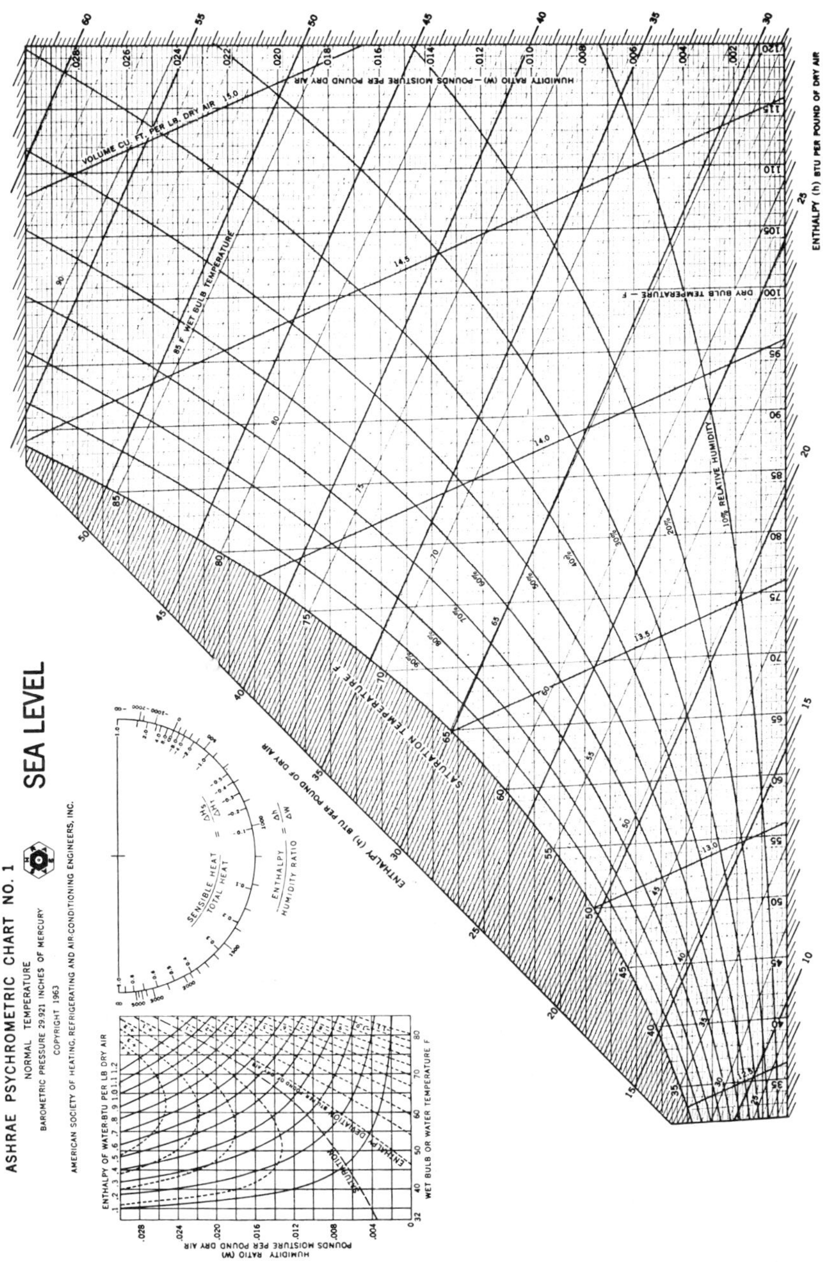

Figure C-27. Psychometric chart. Reprinted by permission of ASHRAE.

Appendix D Conversion Tables

The following tables are included in this section:

Table D-1
Metric-English Conversions

	To Convert From	To	Multiply By
Length			
	ft	m	.3048
	m	ft	3.281
	in	mm	25.4
	mm	in	.0394
Area			
	ft^2	m^2	0.929
	m^2	ft^2	10.76
	in^2	mm^2	645.2
	mm^2	in^2	.00155
Volume			
	ft^3	m^3	.02832
	m^3	ft^3	35.31
	in^3	mm^3	16387
	mm^3	in^3	.000061
	(US) gal	l	3.785
	l	(US) gal	.2642

Table D-1 (continued)

	From	To	Multiply by
Volume flow rate			
	CFM	m^3/hr	1.699
	m^3/hr	CFM	.5886
	$SCFM_{60°F}$	$m^3/hr_{0°C}$	1.607
	$m^3/hr_{0°C}$	$SCFM_{60°F}$	.6223
	$SCFM_{60°F}$	$m^3/hr_{15°C}$	1.696
	$m^3/hr_{15°C}$	$SCFM_{60°F}$	.5896
Mass flow rate			
	lbm/min	kg/hr	27.22
	kg/hr	lbm/min	.03674
	lbm/hr	kg/hr	.4536
	kg/hr	lbm/hr	2.205
Specific volume			
	ft^3/lbm	m^3/kg	.06243
	m^3/kg	ft^3/lbm	16.02
Polytropic or adiabatic head			
	ft-lbf/lbm	N·m/kg	2.989
	N·m/kg	ft-lbf/lbm	.3346
	ft-lbf/lbm	kN·m/kg	.002989
	kN·m/kg	ft-lbf/lbm	334.6
	Btu/lbm	kJ/kg (kN·m/kg)	2.326
	kJ/kg	Btu/lbm	.4299
Pressure			
	psi	bar	.06895
	bar	psi	14.5
	psi	kPa	6.895
	kPa	psi	.145
Power			
	hp	kW	.7457
	kW	hp	1.341

Table D-2 Temperature*

°C ←	Degrees	→ °F	°C ←	Degrees	→ °F	°C ←	Degrees	→ °F	°C ←	Degrees	→ °F	°C ←	Degrees	→ °F	°C ←	Degrees	→ °F	°C ←	Degrees	→ °F
−273.15	−459.7		−26.1	−15	+5.0	8.3	47	116.6	42.2	108	226.4	75.6	168	334.4	108.9	228	442.4	227	440	824
−267.8	−450		−25.6	−14	+6.8	8.9	48	118.4	42.8	109	228.2	76.1	169	336.2	109.4	229	444.2	232	450	842
−262.2	−440		−25.0	−13	+8.6	9.4	49	120.2	43.3	110	230.0	76.7	170	338.0	110.0	230	446.0	238	460	860
−256.7	−430		−24.4	−12	+10.4	10.0	50	122.0	43.9	111	231.8	77.2	171	339.8	110.6	231	447.8	243	470	878
−251.1	−420		−23.9	−11	+12.2	10.6	51	123.8	44.4	112	233.6	77.8	172	341.6	111.1	232	449.6	249	480	896
−245.6	−410		−23.3	−10	+14.0	11.1	52	125.5	45.0	113	235.4	78.3	173	343.4	111.7	233	451.4	254	490	914
−240.0	−400		−22.8	−9	+15.8	11.7	53	127.4	45.6	114	237.2	78.9	174	345.2	112.2	234	453.2	260	500	932
−234.4	−390		−22.2	−8	+17.6	12.2	54	129.2	46.1	115	239.0	79.4	175	347.0	112.8	235	455.0	266	510	950
−228.9	−380		−21.7	−7	+19.4	12.8	55	131.0	46.7	116	240.8	80.0	176	348.8	113.3	236	456.8	271	520	968
−223.3	−370		−21.1	−6	+21.2	13.3	56	132.8	47.2	117	242.6	80.6	177	350.6	113.9	237	458.6	277	530	986
−217.8	−360		−20.6	−5	+23.0	13.9	57	134.6	47.8	118	244.4	81.1	178	352.4	114.4	238	460.4	282	540	1004
−212.2	−350		−20.0	−4	+24.8	14.4	58	136.4	48.3	119	246.2	81.7	179	354.2	115.0	239	462.2	288	550	1022
−206.7	−340		−19.4	−3	+26.6	15.0	59	138.2	48.9	120	248.0	82.2	180	356.0	115.6	240	464.0	293	560	1040
−201.1	−330		−18.9	−2	+28.4	15.6	60	140.0	49.4	121	249.8	82.8	181	357.8	116.1	241	465.8	299	570	1058
−195.6	−320		−18.3	−1	+30.2	16.1	61	141.8	50.0	122	251.6	83.3	182	359.6	116.7	242	467.6	304	580	1076
−190.0	−310		−17.8	0	+32.0	16.7	62	143.6	50.6	123	253.4	83.9	183	361.4	117.2	243	469.4	310	590	1094
−184.4	−300		−17.2	+1	+33.8	17.2	63	145.4	51.1	124	255.2	84.4	184	363.2	117.8	244	471.2	316	600	1112
−178.9	−290		−16.7	+2	+35.6	17.8	64	147.2	51.7	125	257.0	85.0	185	365.0	118.3	245	473.0	321	610	1130
−173.3	−280		−16.1	+3	+37.4	18.3	65	149.0	52.2	126	258.8	85.6	186	366.8	118.9	246	474.8	327	620	1148
−169.5	−273.15	−459.7	−15.6	+4	+39.2	18.9	66	150.8	52.8	127	260.6	86.1	187	368.6	119.4	247	476.6	332	630	1166
−167.8	−270	−454	−15.0	+5	+41.0	19.4	67	152.6	53.3	128	262.4	86.7	188	370.4	120.0	248	478.4	338	640	1184
−162.2	−260	−436	−14.4	+6	+42.8	20.0	68	154.4	53.9	129	264.2	87.2	189	372.2	120.6	249	480.2	343	650	1202
−156.7	−250	−418	−13.9	+7	+44.6	20.6	69	156.2	54.4	130	266.0	87.8	190	374.0	121.1	250	482.0	349	660	1220
−151.1	−240	−400	−13.3	+8	+46.4	21.1	70	158.0	55.0	131	267.8	88.3	191	375.8	122.4	252	485.6	354	670	1238
−145.6	−230	−382	−12.8	+9	+48.2	21.7	71	159.8	55.6	132	269.6	88.9	192	377.6	123.3	254	489.2	360	680	1256
−140.0	−220	−364	−12.2	+10	+50.0	22.2	72	161.6	56.1	133	271.4	89.4	193	379.4	124.4	256	492.8	366	690	1274
−134.4	−210	−346	−11.7	+11	+51.8	22.8	73	163.4	56.7	134	273.2	90.0	194	381.2	125.5	258	496.4	371	700	1292
−128.9	−200	−328	−11.1	+12	+53.6	23.3	74	165.2	57.2	135	275.0	90.6	195	383.0	126.7	260	500.0	377	710	1310
−123.3	−190	−310	−10.6	+13	+55.4	23.9	75	167.0	57.8	136	276.8	91.1	196	384.8	127.8	262	503.6	382	720	1328
−117.8	−180	−292	−10.0	+14	+57.2	24.4	76	168.8	58.3	137	278.6	91.7	197	386.6	128.9	264	507.2	388	730	1346
−112.2	−170	−274	−9.4	+15	+59.0	25.0	77	170.6	58.9	138	280.4	92.2	198	388.4	130.0	266	510.8	393	740	1364
−106.7	−160	−256																		
−101.1	−150	−238																		
−95.6	−140	−220																		
−90.0	−130	−202																		

Table D-2 (continued)

°C ←	Degrees	→ °F
−84.4	−120	−184
−78.9	−110	−166
−73.3	−100	−148
−67.8	−90	−130
−62.2	−80	−112
−56.7	−70	−94
−51.1	−60	−76
−45.6	−50	−58
−40.0	−40	−40.0
−39.4	−39	−38.2
−38.9	−38	−36.4
−38.3	−37	−34.6
−37.8	−36	−32.8
−37.2	−35	−31.0
−36.7	−34	−29.2
−36.1	−33	−27.4
−35.6	−32	−25.6
−35.0	−31	−23.8
−34.4	−30	−22.0
−33.9	−29	−20.2
−33.3	−28	−18.4
−32.8	−27	−16.6
−32.2	−26	−14.8
−31.7	−25	−13.0
−31.1	−24	−11.2
−30.6	−23	−9.4
−30.0	−22	−7.6
−29.4	−21	−5.8
−28.9	−20	−4.0
−28.3	−19	−2.2
−27.8	−18	−0.4
−27.2	−17	+1.4
−26.7	−16	+3.2

°C ←	Degrees	→ °F
−8.9	+16	+60.8
−8.3	+17	+62.6
−7.8	+18	+64.4
−7.2	+19	+66.2
−6.7	+20	+68.0
−6.1	+21	+69.8
−5.5	+22	+71.6
−5.0	+23	+73.4
−4.4	+24	+75.2
−3.9	+25	+77.0
−3.3	+26	+78.8
−2.8	+27	+80.6
−2.2	+28	+82.4
−1.7	+29	+84.2
−1.1	+30	+86.0
−0.6	+31	+87.8
0.0	+32	+89.6
0.6	33	91.4
1.1	34	93.2
1.7	35	95.0
2.2	36	96.8
2.8	37	98.6
3.3	38	100.4
3.9	39	102.2
4.4	40	104.0
5.0	41	105.8
5.5	42	107.6
6.1	43	109.4
6.7	44	111.2
7.2	45	113.0
7.8	46	114.8

°C ←	Degrees	→ °F
25.6	78	172.4
26.1	79	174.2
26.7	80	176.0
27.2	81	177.8
27.8	82	179.6
28.3	83	181.4
28.9	84	183.2
29.4	85	185.0
30.0	86	186.8
30.6	87	188.6
31.1	88	190.4
31.7	89	192.2
32.2	90	194.0
32.8	91	195.8
33.3	92	197.6
33.9	93	199.4
34.4	94	201.2
35.0	95	203.0
35.6	96	204.8
36.1	97	206.6
36.7	98	208.4
37.2	99	210.2
37.8	100	212.0
38.3	101	213.8
38.9	102	215.6
39.4	103	217.4
40.0	104	219.2
40.6	105	221.0
41.1	106	222.8
41.7	107	224.6

°C ←	Degrees	→ °F
59.4	139	282.2
60.0	140	284.0
60.6	141	285.8
61.1	142	287.6
61.7	143	289.4
62.2	144	291.2
62.8	145	293.0
63.3	146	294.8
63.9	147	296.6
64.4	148	298.4
65.0	149	300.2
65.6	150	302.0
66.1	151	303.8
66.7	152	305.6
67.2	153	307.4
67.8	154	309.2
68.3	155	311.0
68.9	156	312.8
69.4	157	314.6
70.0	158	316.4
70.6	159	318.2
71.1	160	320.0
71.7	161	321.8
72.2	162	323.6
72.8	163	325.4
73.3	164	327.2
73.9	165	329.0
74.4	166	330.8
75.0	167	332.6

°C ←	Degrees	→ °F
92.8	199	390.2
93.3	200	392.0
93.9	201	393.8
94.4	202	395.6
95.0	203	397.4
95.6	204	399.2
96.1	205	401.0
96.7	206	402.8
97.2	207	404.6
97.8	208	406.4
98.3	209	408.2
98.9	210	410.0
99.4	211	411.8
100.0	212	413.6
100.6	213	415.4
101.1	214	417.2
101.7	215	419.0
102.2	216	420.8
102.8	217	422.6
103.3	218	424.4
103.9	219	426.2
104.4	220	428.0
105.0	221	429.8
105.6	222	431.6
106.1	223	433.4
106.7	224	435.2
107.2	225	437.0
107.8	226	438.8
108.3	227	440.6

°C ←	Degrees	→ °F
131.3	268	514.4
132.2	270	518.0
133.3	272	521.6
134.4	274	525.2
135.6	276	528.8
136.7	278	532.4
137.8	280	536.0
138.9	282	539.6
140.0	284	543.2
141.1	286	546.8
142.2	288	550.4
143.3	290	554.0
144.4	292	557.6
145.6	294	561.2
146.7	296	564.8
147.8	298	568.4
149	300	572
154	310	590
160	320	608
166	330	626
171	340	644
177	350	662
182	360	680
188	370	698
193	380	716
199	390	734
204	400	752
210	410	770
216	420	788
221	430	806

°C ←	Degrees	→ °F
399	750	1382
404	760	1400
410	770	1418
416	780	1436
421	790	1454
427	800	1472
432	810	1490
438	820	1508
443	830	1526
449	840	1544
454	850	1562
460	860	1580
466	870	1598
471	880	1616
477	890	1634
482	900	1652
488	910	1670
493	920	1688
499	930	1706
504	940	1724
510	950	1742
516	960	1760
521	970	1778
527	980	1796
532	990	1814
538	1000	1832
566	1050	1922
593	1100	2012
621	1150	2102
649	1200	2192
677	1250	2282

* Reprinted by permission of Elliott Company, Jeannette, PA, Bulletin Q-49, 1977.

Note: $°C = \frac{5}{9}(°C - 32)$ $\quad K = °C + 273.15$

$°F = \frac{9}{5}(°C) + 32$ $\quad °R = °F + 459.67$

Bibliography

American Petroleum Institute. *API Standard 617, Centrifugal Compressors for General Refinery Services,* fourth edition, Washington, D.C.: API, November 1979.

Elliott Company. *Compressor Refresher,* Jeannette, PA: Elliott Company, 1975.

Ibid. *Engineers Guide to Modern International Units of Measurement for Rotating Machinery,* Bulletin Q49, Jeanette, PA: Elliott Company, 1975.

Ibid. *Quick Selection Methods for Elliott Multistage Centrifugal Compressors,* Bulletin P-25S-Rev, Jeanette, PA: Elliott Company 1981.

Gas Processors Suppliers Association. *Engineering Data Book,* ninth edition, fifth revision, Tulsa, OK: GPSA, 1981.

Ibid. *SI Engineering Data Book,* Tulsa, OK: GPSA, 1980.

Ingersoll-Rand Company. *Compressed Air and Gas Data,* third edition, Washington, NJ: Ingersoll-Rand Company, 1980.

Keenan, J. H., F. G. Keyes, P. G. Hill, and J. G. Moore. *Steam Tables (English Units),* New York: John Wiley and Sons, Inc., 1969.

Lapina, R. P. "A Quick Approach to Estimating Hydrocarbon K and Z Values for Compressor Applications," *CAGI Technical Digest,* vol. 13, no. 2, Cleveland: Compressed Air and Gas Institute; and *Chemical Engineering* (as "How to Estimate Compressibility Factors and Specific Heat Ratios for Hydrocarbon Gases"), New York: McGraw-Hill Publications, February 8, 1982.

Ibid. "Can You Rerate Your Centrifugal Compressor?", *Chemical Engineering,* New York: McGraw-Hill Publications, January 20, 1975 (originally written for the CAGI Technical Article Program).

Ibid. "Single Stage Centrifugal Compressor Performance Curves—Get More Out of Them," *CAGI Technical Digest,* vol. 9, no. 2., Cleveland: Compressed Air and Gas Institute; and *Chemical Engineering* (as "How to Use the Performance Curves to Evaluate Behavior of Centrifugal Compressors"), New York: McGraw-Hill Publications, January 25, 1982.

Nelson, L. C. and E. F. Obert. "Generalized Compressibility Charts," *Chemical Engineering,* New York: McGraw-Hill Publications, July 1954.

Pratt & Whitney Aircraft, *Aeronautical Vest-Pocket Handbook,* 18th edition, 29th printing, East Hartford, CT: Pratt & Whitney Aircraft, June 1978.

Thermodynamics Research Center Hydrocarbon Project. "Selected Values of Properties of Hydrocarbons and Related Compounds," College Station, TX: Thermodynamics Research Center, Texas A & M University, 1981 (loose-leaf data sheets extant, 1981).

Van Wylen, G. J. and R. E. Sonntag. *Fundamentals of Classical Thermodynamics,* second printing, New York: John Wiley and Sons, Inc., 1968.

Index

Notes

Notes

Notes